$11.00

Builder's Guide To Construction Financing

James E. Newell

Craftsman Book Company
6058 Corte Del Cedro, Carlsbad, CA 92008

Library of Congress Cataloging in Publication Data

Newell, James E

Builder's guide to construction financing.

Includes index.

1. Construction industry--Finance.
2. Mortgage loans. I. Title.

HD9715.A2N44 658.1'522 79-1121

ISBN 0-910460-66-3

Third Printing 1985

Introduction

This book is primarily for builders, developers, property owners, and anyone who buys and sells land and buildings for a profit. It is a guide to loans, lending sources, and investors dealing in loans secured by real estate. It shows how to obtain funds for refinancing, building ventures, leasehold financing, and property improvement. Information is included for those who are unfamiliar with certain aspects of real estate financing as well as the professional who may wish to refer to specific funding procedures as he prepares for a new project. It is also a guide to outright sales financing in the promotion and marketing of the builder's finished product. The real estate financing procedures outlined here include conventional and government insured or guaranteed loans as provided by financial institutions. For although it is possible to have an existing FHA or GI loan on property and still borrow cash conventionally on the equity, the primary purpose of a new government insured or guaranteed loan is to make possible an original low down payment at a lower interest rate, and in some instances to refinance property, *only in the amount of the existing encumbrances.* It is for this reason that government sponsored loans are not used as a refinance device for the purpose of obtaining operating capital for the individual applicant.

That means going to conventional money people who specialize in equity financing, that is, the Consolidator, the Hardmoneyer, the Discounter, and the Hypothecator. Without distinguishing mortgages and trust deeds as an exclusive instrument of any particular state, this manual takes into consideration that many states may well have similar procedures and documents which are used for conveying title or the vesting of interests. Variations prevail. But for the purposes of this book, mortgages and trust deeds are generally used interchangeably.

This manual is not a course of instruction, and it cannot be the final answer to every builder's problems of financing. But it can be a valuable guide to borrowing and lending possibilities. Whether or not you use the services of a mortgage loan broker, this book has tips for "doing it yourself" which may save time and money. Continuous awareness of your borrowing capacity is only good housekeeping. Loans can be shopped for—there is always the possibility of getting a cheaper rate.

Contents

1

Valuation Of Property For Loan Purposes

Commercial lenders make loans against value. These are known as secured loans because property provides security to the lender so that he can recover the amount of the loan. If the borrower defaults, the lender can exercise his rights under a trust deed or mortgage and sell the property for the amount due. Thus, lenders are concerned with the value of the property and its remaining equity if there is already a loan on the property.

Equity is fully dependent on realistic property value. What you may suppose is fair market value is only a guess when you sell outright. This is also true when determining equity in the absence of a sale. However, the "risk rate" can be very flexible from one type of investment to another, and the amount of money placed in escrow for you depends on the risk taken. In other words, the higher LVR, or *loan to value ratio,* the higher the risk to the lender. This book will show you how to finance your building plans and how to conduct an independent loan survey. By surveying the lending activity in your area and the current money policy of the lenders in your survey, you can accurately assess the loan to value risk for your own property. By this survey process you can compare the current loan market and its relationships to your proposed debt finances. Chances are good that a lender will make his own appraisal. But you can assume the position of a salesperson, since what you learn in your survey places you in a position to negotiate with confidence. You must sell the lender on the value of your property by means of the independent loan survey.

Appraised Value Vs. Hypothesis

In order to save time and get your money's worth, it is important to have a good understanding of the loan value of the property you now own or are considering. Since there are governmental controls on the amount of risk the lender may take, his opinion of value depends largely on the information you can furnish him. Opinions of value vary substantially; and in order to shop for the best loan to value on the best available terms, you must sell your opinion. That opinion must be based on your own investigation of the value of the real estate, and will tell you how far you can proceed realistically and with conviction in shopping for the loan.

It is important to the investor in loans secured by real estate to have an informed opinion of the security; and although an investor will make his own appraised value, you also must be able to establish your own opinion with confidence in the negotiation. Without a conviction based on your own investigations, your opinion is only a hypothesis and you leave yourself at a disadvantage in the search for the most generous lender as well as the best loan.

Your Value Estimate

Your valuation must be based on today's market, and should not be a figure projected to market conditions of the future. There are three generally accepted approaches to value: 1.) the cost approach, 2.) the income approach, and 3.) the market approach. For the potential borrower, value estimates for equity funding, finance, refinance or new construction may be limited to the *principle of substitution* rather than those of various appraisal methods applied to income property such as apartment houses, industrial centers and shopping centers. Unless your property is included in an income property category, your mortgage lender will use this principle of substitution as his main determination of value even though he figures in current construction costs.

Appraisals will be discussed in a later chapter as they relate to the loan process. But in order to find your own opinion of value, you must be conversant with appraisal methods even though as a layman you may not be able to speak the appraiser's language. A professional appraiser uses the cost approach on new structures. That is, he must detail what it costs to build the structure using special up-to-date cost tables. He then adds the value of the land to the total cost of today's labor and materials. A variety of depreciation factors are applied. When the appraiser has a semifinal figure for value, he must still make the most important observation—the comparable sales in the area surrounding your property. Comparing recent sales for his principle of substitution serves as an overall check. When appraising an older property he uses the cost approach and builds the structure on paper; but after depreciation factors he then adjusts the figure by applying comparable sales. If the property is a commercial building within which a business can be conducted, he uses a separate capitalized approach to income as well as the cost-to-build approach. The principle of substitution is not much help to him in commercial buildings, but he determines the highest and best use for the property and then applies a depreciation factor. For commercial or income property, the appraiser usually considers a marketing approach to value and applies similar types of business structures in the area as comparables. If he is appraising apartments, he will probably use all three approaches to value.

If you are not able to use the cost approach or the income approach, you may still be able to use the market approach by digging up recent sales of comparable properties in your area. In addition to the professional appraiser's opinion the lender obtains, your own investigation may sell the lender into a more favorable loan than would otherwise be granted. You must be able to point out to him why you believe in your property's worth and why you hold to a certain interest rate opinion. Most important, explain to him why you are inclined to a lower annual percentage rate. You must make the lender believe he has a low risk in loaning on your property. Keep in mind that a mortgage lender assumes that the life of the average building is fifty years, and that the effective life—depending on the property—can be fewer than fifty years. If you can show him that your property has an effective life which is longer than normal, it may determine the length of the loan, or its amortization, and the rate of interest.

Appendix I shows a market sales survey form used by FHA appraisers to make preliminary checks of comparable properties. Information gathered in your sales survey can be consolidated into the one complete "estimate of market price comparison."

To estimate the fair market value of your property by using the "principle of substitution," first determine the highest and best use of land you plan to borrow on. Establish in general where your property falls in approximate land use in relation to the area. If there is a similarity in the construction and design of other properties in the area—as in the case of tracts or developments—establish the "building value" (including the land) rather than construction costs alone as a standard by which to estimate your own property value. Find the most recent sale prices of similar property.

New owners or recent buyers are often more willing to disclose property prices than sellers. Sellers are usually more reluctant to discuss the conditions of sale and the true price of the property. A new owner, on the other hand, would have no reason to hide the sale price if he is tactfully approached. By keeping abreast of recent sales in the area, and getting acquainted with new owners, you can take a survey that cannot be equalled by any other valuation approach. In the survey try to record just-sold property with approximately the same square footage and same overall characteristics as your property.

Contact an agent involved in the sale of property in the area, being careful to give the impression that you are just checking the market prices for a possible sale of your property in the future. But beware of the agent's opinion as to the value of your property. He may tell you what he believes you want to hear: an inflated value estimate.

There are other ways to check value, and some of them are

outlined below. But all should be correlated with the values found in your independent survey of area sales.

Checking Recorded Deeds

Experienced appraisers and brokers sometimes check government tax stamps on recorded deeds in the county records office. This is a good start for an investigation. However, remember that tax stamps may not reflect the actual price of the property you are investigating, because the tax is not applicable to existing mortgages or trust deeds taken over, or "assumed," by buyers in subsequent transactions. Revenue stamps are issued in the amount of $1.10 per $1,000.00 and $.55 per $500.00 or less on the *cash only*. For example, on a total of $9.35 in revenue stamps you could estimate the cash paid at $8,000—$8,500. This amount may have been the down payment.

If the exact amount of the existing loans recorded on a recent purchase is known, a rough calculation of value can be made. For instance, if the amount of the old (or existing) mortgage is $35,000, and the probable second mortgage carried by the seller is 15%, the price covered was $52,000 as shown below.

1st loan	$35,000 (old mortgage or assumption of existing loan)
2nd loan	$8,500 (carried back by seller in lieu of cash)
Cash down payment	$8,500 (tax stamps showing $9.35)

If the tax stamps totalled $14.30, say, it may reflect that the cash down payment was $12,750 with a corresponding reduction in the second loan by $4,250.

This breakdown is typical of sales in many states where the seller raises his price to offset a discounted trust deed or mortgage later on. These figures may vary depending on individual findings. Tax stamps can support a market survey of sales only if considerable information is already known before the visit to the recorder's office.

An Outside Opinion of Value

Any owner of improved residential property can request a

VETERANS ADMINISTRATION

REQUEST FOR DETERMINATION OF REASONABLE VALUE (Real Estate)

CASE NUMBER

On receipt of "Certificate of Reasonable Value" or advice from the Veterans Administration that a "Certificate of Reasonable Value" will not be issued, we agree to forward to the appraiser the approved fee which we are holding for this purpose.

1. STATUS OF PROPERTY
☐ A. PROPOSED ☐ B. EXISTING, NOT PREVIOUSLY OCCUPIED ☐ C. EXISTING, PREVIOUSLY OCCUPIED ☐ D. ALTERATIONS, IMPROVEM'TS. OR REPAIRS ☐ E. REFINANCING-VETERAN APPLICANT OWNS AND OCCUPIES RESIDENCE AS HOME

2. CONSTRUCTION COMPLETED BEFORE DATE HEREOF
☐ A. WITHIN 12 CALENDAR MOS. ☐ B. MORE THAN 12 CALENDAR MOS.

3. NAME AND ADDRESS OF FIRM OR PERSON MAKING REQUEST *(Complete mailing address. Include ZIP Code.)*

4. PROPERTY ADDRESS *(Include ZIP Code)*

5. TYPE OF PROPERTY
☐ HOME
☐ MOBILE HOME LOT

6. MANDATORY HOME ASSOCIATION MEMBERSHIP?
☐ YES ☐ NO

7A. NO. BLDGS.

7B. NO. LIVING UNITS

8. LOT DIMENSIONS

9. DESCRIPTION											
DETACHED	WOOD SIDING	CINDER BLOCK	SPLIT LEVEL	NO. ROOMS	DINING ROOM	CAR GARAGE	GAS	CEN. AIR COND.
SEMI-DET.	WOOD SHINGLE	STONE	% BASEMENT	BEDROOMS	KITCHEN	CAR CARPORT	UNDERGRD.WIRE	TYPE HEAT. & FUEL
ROW	ALUM. SIDING	BRICK & BLOCK	SLAB	BATHS	FAMILY RM.	WATER *(Public)*	SEWER *(Public)*	
CONDOMINIUM	ASB. SHINGLE	STUCCO	CRAWL SPACE	½ BATHS	UTILITY RM.	WATER *(Comm.)*	SEWER *(Comm.)*	ROOFING DESCRIP.
	BRICK VENEER	STORIES	YRS. EST. AGE	LIVING RMS.	FIREPLACE	WATER *(Ind.)*	SEPTIC TANK	

10. LEGAL DESCRIPTION

11. TITLE LIMITATIONS, INCLUDING EASEMENTS, RESTRICTIONS, ENCROACHMENTS, HOMEOWNERS ASSOCIATION AND SPECIAL ASSESSMENTS, ETC.

12. OFFSITE IMPROVEMENTS
A. STREET SURFACE:
B. STREET ACCESS ☐ PRIV. ☐ PUB.
C. STREET MAINT. ☐ PRIV. ☐ PUB.
D. ADD'L. IMPROVEMENTS ☐ STORM SEWER ☐ SIDEWALK ☐ CURB/GUTTER

13. VETERAN PURCHASER'S NAME AND ADDRESS *(Complete mailing address. Include ZIP Code)*

14. REMOVABLE EQUIPMENT INCLUDED IN PURCHASE PRICE OR COST
☐ RANGE OR COUNTER TOP UNIT ☐ DISHWASHER ☐ REFRIGERATOR
☐ AUTOMATIC WASHER ☐ DRYER ☐ WALL-TO-WALL CARPETING
☐ OTHER(S) *(Specify)*

15A. OCCUPANT'S NAME | 15B. TELEPHONE NO. | 16A. BROKER'S NAME | 16B. TELEPHONE NO.

17. DATE AND TIME AVAILABLE FOR INSPECTION AM PM | 18. KEYS AT *(Address)* | 19. NAME OF OWNER

20. COMPLIANCE INSPECTIONS WILL BE OR WERE MADE BY ☐ FHA ☐ VA ☐ NONE MADE

21. NUMBER OF MASTER CERTIFICATE OF REASONABLE VALUE *(If any)*

22. PROPOSED SALES CONTRACT ATTACHED ☐ YES ☐ NO

23. CONTRACT NO. PREVIOUSLY APPROVED BY VA THAT WILL BE USED

24A. NAME AND ADDRESS OF BUILDER *(Include ZIP Code)* | 24B. TELEPHONE NO. | 25A. NAME AND ADDRESS OF WARRANTOR *(Include ZIP Code)* | 25B. TELEPHONE NO.

26. PLANS *(Check one)* ☐ FIRST SUBMISSION ☐ REPEAT CASE *(If repeat case, complete Item 27)*

27. PLANS PREVIOUSLY PROCESSED UNDER VA CASE NO.

28. ANNUAL REAL EST. TAXES *(If exist. construction)* $

29. COMMENTS ON SPECIAL ASSESSMENTS AND/OR HOMEOWNER ASSOCIATION CHARGES

30. SHOW BELOW: Shape, location, distance from nearest intersection, and street names. Mark N at north point.

EQUAL OPPORTUNITY IN HOUSING - NOTICE

Federal laws and regulations prohibit discrimination because of race, color, religion, national origin, or sex in the sale or rental or financing of residential property. Numerous state statutes and local ordinances also prohibit such discrimination.

Non-compliance with applicable antidiscrimination laws and regulations in respect to any property included in this request shall be a proper basis for refusal by the VA to do business with the violator and for refusal to appraise properties with which the violator is identified. Denial of participation in any program administered by the Federal Housing Administration because of such violation shall constitute basis for similar action by the VA.

CERTIFICATION REQUIRED ON CONSTRUCTION UNDER FHA SUPERVISION *(Strike out inappropriate phrases in parentheses)*

I hereby certify that plans and specifications and related exhibits, including acceptable FHA Change Orders, if any, supplied to VA in this case, are identical to those (submitted to) (to be submitted to) (approved by) FHA, and that FHA inspections (have been) (will be) made pursuant to FHA approval for mortgage insurance on the basis of proposed construction under Sec.

31A. NAME AND ADDRESS OF PROSPECTIVE LENDER *(Include ZIP Code)* | 31B. TELEPHONE NO. OF LENDER/REQUESTER | 32. SALE PRICE OF PROPERTY $ | 33. REFINANCING AMT. OF PROPOSED LOAN $

34. SIGNATURE OF PERSON AUTHORIZING THIS REQUEST | 35. TITLE | 36. DATE

Federal statutes provide severe penalties for any fraud, intentional misrepresentation, or criminal connivance or conspiracy purposed to influence the issuance of any guaranty or insurance or the granting of any loan by the Administrator.

37. DATE OF ASSIGNMENT | 38. NAME OF APPRAISER

Figure 1-1

GI appraisal through the Veteran's Administration. Ostensibly, this is for the purpose of obtaining a commitment for a possible veteran prospective buyer. Although the information is for your own purposes, if you later sell the property, the Certificate of Reasonable Value issued by the Administration can be a convenient marketing instrument.

GI loans placed on property being sold are 100% loan to value on low to moderately priced homes. However, appraisals are not necessarily made for fair market value. You request this one to determine if the fair market value reaches the amount of the loan requested. You should ask that the appraisal be for a value greater than your estimation of the value of the property. The GI appraiser then conducts a full appraisal and the Certificate of Reasonable Value (Figure 1-2) shows a loan commitment for less than you had requested.

For example, if you have generally assumed the value of your property to be $50,000 and you request the VA to appraise it for a $57,000 loan commitment, a market survey will indicate that the 100% GI loan would be guaranteed for around $50,000, or a fair market value. On the other hand, if you understate the value by requesting an appraisal for a $47,000 loan commitment and the value is actually $57,000, the Certificate of Reasonable Value will show a 100% loan of $47,000—not a full appraised value.

A Certificate of Reasonable Value Request Form (Figure 1-1) and the Certificate itself can be obtained for $70.00 by contacting the real estate loan section of the VA in your area.

The Tax Assessor's Opinion

Still another check can be made from the most recent tax assessor's appraisal. Many assessors make a valuation at about 80% of fair market value. In this way the tax paid by property owners is seldom disputed, and the taxpayers choose not to argue with the assessor when they notice how low the appraiser has determined the value to be. Since most owners do not realize this has been done for all properties in the area, very little is said for fear the assessor will notice the "error" and correct the appraisal to the true value—thereby increasing the assessment and the taxes.

To determine the number of hundreds of dollars of

The Reasonable Value as set forth herein is predicated upon conditions recited below as may be applicable.

VETERANS ADMINISTRATION

CERTIFICATE OF REASONABLE VALUE

CASE NUMBER

1. ESTABLISHED REASONABLE VALUE OF PROPERTY $

2. REMAINING ECONOMIC LIFE OF PROPERTY IS ESTIMATED TO BE NOT LESS THAN ____ YEARS

3. BUILDING SIZE (Check and enter no.) ☐ CUBIC ☐ SQUARE ____ FT.

4. EXPIRATION OF VALIDITY PERIOD

5. STATUS OF PROPERTY ☐ A. PROPOSED ☐ B. EXISTING, NOT PREVIOUSLY OCCUPIED ☐ C. EXISTING, PREVIOUSLY OCCUPIED ☐ D. ALTERATIONS, IMPROVEM'TS. OR REPAIRS ☐ E. REFINANCING–RESIDENCE OWNED AND OCCUPIED BY VETERAN APPLICANT AS HIS HOME

6. CONSTRUCTION COMPLETED BEFORE DATE HEREOF ☐ A. WITHIN 12 CALENDAR MOS. ☐ B. MORE THAN 12 CALENDAR MOS.

7. NAME AND ADDRESS OF FIRM OR PERSON MAKING REQUEST (Complete mailing address. Include ZIP Code)

8. PROPERTY ADDRESS (Include ZIP Code)

9. TYPE OF PROPERTY ☐ HOME ☐ FARM ☐ BUSINESS

10. NO. BLDGS.

11. NO. LIVING UNITS

12. LOT DIMENSIONS

13. DESCRIPTION											
	WOOD SIDING		CINDER BLOCK	SPLIT LEVEL		NO. ROOMS		DINING ROOM	CAR GARAGE	GAS	CEN. AIR COND.
DETACHED	WOOD SHINGLE		STONE	% BASEMENT		BEDROOMS		KITCHEN	CAR CARPORT	UNDERGRD. WIRE	TYPE HEAT. & FUEL
SEMI-DET.	ALUM. SIDING		BRICK & BLOCK	SLAB		BATHS		FAMILY RM.	WATER (Public)	SEWER (Public)	
ROW	ASB. SHINGLE		STUCCO	CRAWL SPACE		1/2 BATHS		UTILITY RM.	WATER (Comm.)	SEWER (Comm.)	ROOFING DESCRIP.
CONDOMINIUM	BRICK VENEER		STORIES	YRS. EST. AGE		LIVING RM.		FIREPLACE	WATER (Ind.)	SEPTIC TANK	

14. LEGAL DESCRIPTION

15. TITLE IS IN FEE SIMPLE, FREE OF ALL ENCROACHMENTS, EASEMENTS AND OTHER LIMITATIONS WITH THE EXCEPTION OF THE FOLLOWING

16. TYPE OF STREET PAVING — CURB, SIDEWALK, STORM SEWER

17. VETERAN PURCHASER'S NAME AND ADDRESS (Complete mailing address. Include ZIP Code)

18. REMOVABLE EQUIPMENT IN VALUE ☐ RANGE OR COUNTER TOP UNIT ☐ DISHWASHER ☐ REFRIGERATOR ☐ AUTOMATIC WASHER ☐ DRYER ☐ WALL-TO-WALL CARPETING ☐ OTHER(S) (Specify)

GENERAL CONDITIONS

(NOTE THE VETERANS ADMINISTRATION DOES NOT ASSUME ANY RESPONSIBILITY FOR THE CONDITION OF THE PROPERTY. THE CORRECTION OF ANY DEFECTS NOW EXISTING OR THAT MAY DEVELOP WILL BE THE RESPONSIBILITY OF THE PURCHASER.)

1. This certificate will remain effective as to any written contract of sale entered into by an eligible veteran within the validity period indicated.
2. This dwelling conforms with the Minimum Property Requirements prescribed by the Administrator of Veterans Affairs.
3. The aggregate of any loan secured by this property plus the amount of any assessment consequent on any special improvements as to which a lien or right to a lien shall exist against the property, except as provided in Item 29 below, may not exceed the reasonable value in Item 1 above.
4. Proposed construction shall be completed in accordance with the plans and specifications identified below, relating to both on-site and off-site improvements upon which this valuation is based and shall otherwise conform fully to the VA Minimum Property Requirements. Satisfactory completion must be evidenced by either
   A. VA Final Compliance Inspection Report (VA Form 26-1839), or
   B. VA Acceptance of FHA Compliance Inspection Reports (FHA Forms 2051) or other evidence of completion under FHA supervision applicable to proposed construction.
5. By contracting to sell property, as proposed construction or existing construction not previously occupied, to a veteran purchaser who is to be assisted in the purchase by a loan made, guaranteed, or insured by VA, the builder or other seller agrees to place any down payment received by the seller or agent of the seller in a special trust account as required by section 1806 of Title 38, U.S. Code.

SPECIFIC CONDITIONS (Applicable when checked or completed)

19. THE REASONABLE VALUE ESTABLISHED HEREIN FOR THE RELATED PROPERTY IS ☐ BASED UPON OBSERVATION OF THE PROPERTY IN ITS "AS IS" CONDITION ☐ PREDICATED UPON COMPLETION OF REPAIRS LISTED IN ITEM 22 ☐ PREDICATED UPON COMPLETION OF PROPOSED CONSTRUCTION (If checked complete item 20)

20. PROPOSED CONSTRUCTION TO BE COMPLETED (Identify plans, specifications and exhibits)

21. INSPECTIONS REQUIRED ☐ FHA COMPLIANCE INSPECTIONS FOR PROPOSED CONSTRUCTION ☐ VA COMPLIANCE INSPECTIONS ☐ LENDER TO CERTIFY

22. REPAIRS TO BE COMPLETED

23. NAME OF COMPLIANCE INSPECTOR

24. ☐ HEALTH AUTHORITY APPROVAL – Execution of VA Form 26-6395 by the Health Authority indicating approval of the water supply and/or sewage disposal installation is required. (Approval by letter or Health Authority Form may be used.)

25. ☐ This document is subject to the provisions of Executive Orders 11246 and 11375, and the Rules and Regulations of the Secretary of Labor in effect this date, and VA Regulation 4390 through 4393, and also the provision of the certification executed by the builder, sponsor or developer named herein which is on file in this office.

26. ☐ TERMITE CERTIFICATE – The seller shall furnish the veteran-purchaser at no cost to the veteran prior to settlement a written statement (or certification) from a recognized exterminator that based on careful visual inspection of accessible areas and on sounding of accessible structural members, there is no evidence of termite or other wood-destroying insect infestation in the subject property, and, if such infestation previously existed, it has been corrected and any damage due to such infestation has also been corrected or alternatively been fully disclosed as follows . . . .

27. WARRANTY ☐ (If checked, complete Item 28)

28. NAME OF WARRANTOR

29. SEE GENERAL CONDITIONS ABOVE

30. OTHER REQUIREMENTS

31. FLOOD INSURANCE (Check if this property is located in a special Flood Hazard area and flood insurance will be required in accordance with VA Reg. 4326). ☐

32. DATE

33. ADMINISTRATOR OF VETERANS AFFAIRS, BY (Signature of authorized agent)

34. VA OFFICE

Figure 1-2

assessed valuation, take the tax rate per $100 of assessed valuation for your area and divide it into the annual tax bill before exemptions. Since the most common method of taxing is based on 25% of the appraiser's valuation, take the above figure and move the decimal point to the right two places. Then multiply by four to get 80% of the fair market value. Dividing this figure by 80 will give 100% of market value on the day of appraisal. For accuracy, make sure you use the most recent assessor's survey.

2

The Consolidation Loan

Consolidation is a way of obtaining cash in a lump sum from equity. This can be used for reinvestment, or to make a substantial cash purchase to avoid carrying any interest charges. A consolidation lender can cover all your existing encumbrances, plus cash, in one new first trust deed loan. The cash may be invested in income property or used for a building project; it may purchase a discounted security in real estate, or finance home improvements. You may have encumbered your property with a junior mortgage or trust deed, thereby reducing your equity. It may still be possible to cover both the old first and the existing second loan, depending upon the property's loan to value ratio. Many property owners have consolidated when faced with a junior mortgage that comes due with a *balloon* payment after the installment period is terminated. When the final installment payment on a note is greater than the amount of each previous installment payment and the note is due in full, such a balloon can sometimes be renegotiated with the former owner of the property. This is sometimes simpler than it seems at first. This type of financing could have been agreed upon because of the low down payment made to the seller. The price might have been inflated. The seller could have opted to discount the "purchase money instrument" to an investor. Or he may have decided to retain ownership in it to supplement other income. After all, you might point out to him, he has made money on holding it. With an equity in your property of nearly 50% you can consider a complete new hard money loan transaction to pay off the purchase money mortgage. Or you can increase your open-end mortgage, if you have one, or consolidate in the form of a new first loan with a higher interest. *But a consolidation loan is not for you if the total of all existing loan balances, plus other outside obligations to be paid off, is more than 75% of the estimated fair market value.*

It may be possible to obtain the amount of cash you need by simply increasing the existing mortgage. There are some service costs, but it is less expensive than consolidation, provided the increased amount is sufficient. You deal directly with the lender, just as you deal directly with the other sources of funds covered in this book, rather than with an agent.

Loan to Value Ratio

Remember that a 75% LVR is close to an accepted limit of risk in lending circles, even if prime properties are consolidated. And what is considered prime property by one lender may not be so by others. Note that the 75% LVR is for refinancing, not actual sales. Conventional loans for new purchases can be as high as 90%; but a standard LVR has been 80% for some years with some variation from one source to another. Each institution has its own policies which may affect the loan. For example, one institution may follow the *interest rate increase* system in refinancing the same property a second time, rather than charge percentage points for making a new loan. This can be ½% more interest than a loan made for the purchase of property. Government guaranteed or insured loans for refinancing are not suitable for equity fund production. A government loan is for purchasing a home and is not a vehicle for obtaining cash. In some instances a GI or FHA loan can be used to refinance only the amount of the encumbrance on the property; but for complete use of equity these loans are not the best possible. And while a GI loan has no prepayment or payoff fee, an FHA loan may call for some penalty.

Many real estate loan brokers—and in some areas the agents involved in sales — receive their fees from "finding" a loan. In this case the laws state that their status must be revealed and full disclosure made to you *if a separate fee is paid from the loan*. This requirement allows agents for lenders and title companies, attorneys or others, to perform a service such as finding a borrower and to receive compensation for their work. The law prohibits lending institutions from secretly agreeing to return part of the lenders' fee gotten from you to a third referring party in order to obtain business from him. The danger is that some settlement fees can be inflated to cover additional payment for this party, resulting in a higher cost to you for the loan.

Comparing Lender Costs

If a lender is willing to reduce his fees for such items as loan origination, discount points, and other one-time settlement charges, he may gain it back if he charges a higher mortgage interest rate. There is a rule of thumb you can use to calculate the combined effect of the interest rate on your loan and the one-time settlement charges paid by you. While not perfectly accurate, it is usually close enough for meaningful comparisons between lenders. That is, one-time settlement charges equalling 1% of the loan amount increase the interest charge by 1/8%. The 1/8 factor corresponds to a pay-back period of approximately 15 years. If you intend to hold the property only five more years and then sell or pay off the loan, the factor increases to ¼ %.

Here is an example of the rule. Consider only those charges that differ between the lenders. Suppose $30,000 was a sufficient amount of money to borrow as the total cash proceeds to pay off existing encumbrances and still have enough left to carry out your cash plans.

Lender X will make the loan at 8.5% interest. But he charges a 2% origination fee, a $150 application fee, and requires that you use a lawyer for title work at a fee of $300.

Lender Y will make the loan at 9%, but has no additional requirements or charges. As part of the 9% he will not charge an application fee and will absorb the lawyer's fee.

What are the actual charges in each case? Begin by

converting all of Lender X's one-time charges to percentages of the $30,000 loan amount.

2% origination fee	= 2.0% of loan amount
$150 application fee	= 0.5% of loan amount
$300 lawyer's fee	= 1.0% of loan amount
Total	= 3.5% of loan amount

Since each 1% of the loan amount in charges is equivalent to 1/8% increase in interest, the effective interest rate from Lender X is the quoted or contract interest rate, 8.5%, plus .44% (3.5 x 1/8). This is a total of 8.94% interest; since Lender Y has offered a 9% interest rate, Lender X has made a more attractive offer. It is still more attractive if you pay Lender X's one-time charges in cash, thereby taking advantage of a discount.

This rule of thumb is variable according to the length of time you anticipate owning the property before paying off the mortage. As indicated above, the factor increases to ¼% if you expect to pay off in five years. So the effective interest rate for Lender X, given the ¼ factor instead, would be 8.5% plus .87% (3.5 x ¼) for a total of 9.37% interest. Lender X's offer is no longer more attractive than Lender Y's, which was 9%.

In using this rule of thumb be careful of which one-time fees you include in the calculation. For example, if Lender Y did not include a legal fee in his charge but told you that you had to secure legal services in order to obtain the loan from him, you would have to add the legal fee to Lender Y's interest rate.

Annual percentage rate, in this case, means interest on the loan plus charges converted to percentages. You can use this method to compare the effective interest rates of any number of lenders. If they have provided truth-in-lending disclosures as required by law, you are in an even better position to compare. But be sure you have learned of all the charges the lender intends to make. The "good faith estimate" you receive when you make a loan application is a good checklist for this information, but it is not precise. It may not include how the fee and charges are computed. Lenders are required to provide you with a truth-in-lending statement by the time of loan consummation. This is information you should know because it will reveal the annual percentage rate or the *effective* interest rate. The *contract* interest rate is the quoted or advertised interest rate and is not useful in lending comparisons because it includes only the per annum interest rate. The annual percentage rate includes discount points, fees and finance charges. The only problem is that you may not receive advance information in ample time to compare the lender's overall finance charges with those of other lenders. It is true that you will receive a good faith estimate at the time of application; but since the lender is not required to furnish a truth in lending statement at this time, you may or may not get the annual percentage rate information. Since the annual percentage rate the lender charges you on your loan is the most important item of information with which you compare as you shop, you may have to *request* its disclosure at the time of preliminary application. Depending upon many factors including the property appraisal and credit report as well as the supply of funds and the policy of the particular lender, this vital information may not be available at the time of your early contact with a lender. Sometimes money supply or competition play a part. When money is tight a loan is not considered an important lender function, and information is harder to get. At other times actual commitments, although known as preliminary commitments, can be obtained over the phone with the lender "firming up" the commitment made at the time of the preliminary application. Sometimes a hasty, or "windshield," appraisal is made in this way to speed up the process. Of course, the formal appraisal in this case would be made later during processing.

The Truth in Lending Act

As a borrower, you have certain rights under the Truth in Lending Act provisions of the Consumer Credit Protection Act of 1968 (Public Law 90-321; Title 15, U. S. Code 1601 et seq.). The Act requires that borrowers in "consumer credit transactions" be vested with these rights and protections, and receive specified written information from their lenders. The disclosures must be made before credit is extended and before the borrower becomes obligated in connection with the transaction—that is, before the note or mortgage is executed. Among the required disclosures are:

1. The amount of credit a borrower will have for his actual use (*the Amount Financed*); and
2. The *Finance Charge* (consisting primarily of interest but also other fees and charges) expressed both as a dollar amount and as an *Annual Percentage Rate.*

In addition, the act enables a borrower, within three days following the loan transaction, to rescind the transaction, if the loan is secured by a lien on the borrower's residence.

The Board of Governors of the Federal Reserve System has made regulations implementing and interpreting the act. These are entitled "Regulation Z (12 CFR 226)." The text of Regulation Z, and the Truth in Lending Act itself, are presented in an informative pamphlet, "What You Ought To Know About Federal Reserve Regulation Z, Truth in Lending, Consumer Credit Cost Disclosure." It can be obtained from any Federal Reserve Bank or from the Board of Governors of the Federal Reserve System, Washington, D.C. 20551. As indicated in previous pages, the basic interest rate is rarely the only charge. Service charges or carrying charges or *any* charge of any kind on any single transaction, in addition to interest, must now be totalled under the Truth in Lending Law. This total sum is the *finance charge* to be listed as the annual percentage rate of the total charge for the credit transaction.

Using a Loan Broker

Depending on how much time and effort you can spend on seeking a loan, or the distances between you and lender services, it might be cheaper to farm out your loan to an agent. Determine if the loan broker's fee is covered in the lender's origination fee, and if not, add it to the cost of the loan. In any case you must determine whether the encumbrances you are buying will pay off all your existing encumbrances—including a possible second mortgage, an additional third junior lien, delinquent taxes, mechanic's liens, attachments or judgments—and still leave adequate cash left over for investments or purchases. Whether you use an agent or do the work yourself, your first consideration is to get the most money in the form of cash proceeds at the lowest rate of interest with the lowest fees.

Here are some terms you should know in dealing with any lender. *Interest rate,* or the price of money, is by far the most important item affecting monthly installments. It is negotiable with the lender, as is annual percentage rate. *Principal,* the basic amount you pay on, is not necessarily the amount you get in the proceeds. It is based on the value of your property. *Points,* or one of the service charges added to the settlement fees, are a percent (not an interest rate) of the principal amount you are borrowing. One point equals one percent. Next to interest rate, points are most important in assessing a loan bargain. *Annual percentage rate* means interest on the loan plus all charges.

Interest is paid on the remaining principal balance owed each month. Each monthly payment finds a smaller interest due; however, the monthly payments remain constant so a larger portion of the monthly installment is applied on principal—thus reducing the size of the loan. As the months pass and the interest charges become less and less, your loan becomes increasingly smaller. This is called *amortization*—a standard term used in nearly all mortgage loans. The rate of interest should be constant over the full period of amortized loans. There are recent innovations among lenders, however, which call for variable interest rates in some types of repayment procedures; but for purposes of refinance it is doubtful that you will encounter variable interest except as an optional item. Though it is not common now, variable interest rates may appear in the future in permanent mortgages for new projects.

There is nearly always a pay-off penalty when replacing or paying off an old loan. A service charge for processing and paperwork is made for discontinuing an existing loan, and this should be in your original contract. A rule of thumb in this case may be the equivalent of six months' interest on the remaining principal balance. The penalty charged varies among lenders. Check with your loan officer ahead of time, and figure the payoff into the cost of the consolidation. And to get a loan cost analysis, drop in to any savings and loan institution and pick up the booklet "Settlement Costs and You." This is a HUD

(Department of Housing and Urban Development) guide available to the public through state and federal chartered savings and loans.

It is illegal for an agent or a lender to charge or accept a fee where no service has been performed. But this requirement does not prohibit payments to agents for lenders actually performing a service in connection with obtaining a mortgage loan for you which may or may not materialize. An applicant can incur expenses to the lending institution by not completing a loan after the final application has been processed. At what point the lender will begin charging for these services is a matter of business policy. If a service has been performed by an authorized agent who is on an approved list of the institution, he is then on the staff of that institution. He is its *representative.* Make sure that your closing costs or escrow expenses do not show an extra point or fee as a separate item of expense to come out of the proceeds of the loan and to go to a specific party for services that you have not approved or even know about. But it is perfectly legal—even customary, in some cases—to charge the applicant for a service in connection with the loan as long as full disclosure has been made to the applicant. The time to find out about such charges is upon original application rather than when they have begun to build up unknown to you.

An Example of Consolidation

We can use the property we investigated for value in Chapter 1. It was checked in the market survey at the recorder's office and the tax stamp total was found to be $14.30. The buyers had assumed the old first loan of $35,000 with a 25% down payment of $12,750, but after five years the first loan is paid down to $24,500. They have made installment payments since 1972 on the loan at $335.52 with principal and interest which had already been an encumbrance on the property at the rate of 6½% interest per 20 year amortization. Our owner had bought the property when the original $45,000 loan had been paid down to $35,000, and had assumed that loan. The principal amount of the second mortgage was $4,250 and it had been written for a five year term at 10% interest. Payments have been $42.50 monthly including principal and interest on each installment. Although the owner could have cleared up this junior lien during the five year term, he found it advantageous to use his money elsewhere. Now, at the end of five years, he anticipates a balloon payment as a lump sum of $3,614. (The formula for paying off balloon payments is in Appendix IV.) Since he must pay in the near future, he plans a complete refinance plan.

Since a complete pay-off is involved on the $24,500 existing first loan, the owner must anticipate the "six months' interest" rule as a pay-off penalty, as shown below.

$$24{,}500 \times .065 \times 180/360 = 816 \text{ or } \$25{,}316$$

The remaining principal balance on these two loans now looks like this.

$25,316 to pay off old loan
$ 3,614 balloon payment due on 2nd loan
$28,930 Total

Although the owner of this property realizes that the total of $28,930 plus normal administrative costs—reconveyance deed included—is all that is required to clear the loan encumbrances, he wishes to get the maximum amount available on the market in a new first trust deed or mortgage. He can pay off the above loans and keep the difference in cash. After a survey of sales in the area, the owner concludes that the property has increased in value at least as much as comparable property he used in the survey. And the principle of substitution reveals a survey period from 1965 through 1976 showed an increase of 100% in his as well as other areas. Having bought the property in 1972, he used an average figure of 19.35% per year general price increase as an adjustment to comparable sales figures he found. So the property he purchased five years previously is worth approximately $104,000 in 1977.

The owner makes a loan application for an 80% LVR, or $83,200, which is the highest loan commitment he can find in his survey of lending institutions as shown at the right in the Sample Lender Survey.

SAMPLE LENDER SURVEY

Column 1

1977 Appraised Value $104,000
Commitments for refinance 1977

	Lender A	Lender B	Lender C	Lender D
Loan amount	$75,300	$78,000	$83,200	$80,000
Interest rate	9½%	9%	8½%	9½%
Annual percentage rate	9.75%	9.43%	9.13%	9.94%
Length of loan	20 years	25 years	25 years	20 years
Payments per month	$701.92	$654.58	$670.00	$745.72
Loan fee (points)	2%	3.5%	5.0%	3.5%

Column 2

1972 Purchase price $52,000
Loan Commitments to buyer in 1972

	Lender A	Lender B	Lender C	Lender D
Loan amount	up to 80% of sell price	$40,400	Up to 80% of sell price	$41,000
Interest rate	8.4%	7.8%	8%	8%
Annual percentage rate	9.15%	8.5%	8.5%	8.63%
Length of loan	30 years	22 years	25 years	20 years
Payments per month		$320.00		
Loan fee (points)	3.0%	3.0%	2.0%	2.5%

*Note that "payments per month" in Column 2 is omitted for individual study.
Refer to Table of Factors in Appendix III.

Lender C is the most generous in funding and agrees that the property is worth $104,000. But it requires four points in addition to the regular charges. This totals 5% of the loan.

4 points origination fee	=	4.0% of loan amount
$416 application fee	=	0.5% of loan amount
$416 lawyer's fee	=	0.5% of loan amount
Total	=	5.0% of loan amount

The lender makes the preliminary commitment of $83,200 at 8½% interest. An appraisal by the association will verify the value, and the entire application will be presented to the loan committee. The loan's two features—the dollar amount and the interest rate—are advantages that outweigh the high origination fees.

The lowest annual percentage rate available will recover the extra costs of the loan in a comparatively short time. This annual percentage rate can be calculated by adding the quoted or contract interest rate — 8.5% — to the number of points converted to percents and multiplied by the 1/8 factor — .63% — for a total of 9.13%. (Do not confuse this with the amortization rate, which is still 8.5%.) This APR is then calculated on the costs of obtaining the $83,200 loan, as shown in the following.

1. Charges for making new first loan:

4% origination fee	$3,328
Application fee	$ 416
Lawyer's fee	$ 416
Total charge for making loan	$4,160

2. Advances of new first loan:
   $83,200 Principle amount
   -$ 4,160 Settlement charges
   $79,040 Amount advanced

3. Paying off the old first loan:
   $24,500 Remaining principle balance
   $ 816 Prepayment penalty, 6 months' interest
   $25,316 Total obligation on old first loan

4. Proceeds after pay-off:
   $79,040 Amount advanced
   -$25,316 Original loan pay-off
   $53,724 Proceeds before consolidation

The owner-applicant for the new first trust deed loan receives the final figure from Lender C's escrow department. The demands required are the pay-off of the purchase money second trust deed or mortgage, an adjustment of the tax impounds, insurance and other services the lender anticipates are needed. These services, fairly standard in the industry, are nevertheless open to negotiation. They are administrative fees for such things as reconveyances, escrow, title policy and drawing the deed.

The final disbursement of funds from the new loan is complete:

Consolidation

Pay-off of original first loan	$25,316
Charges for making new first loan	4,160
Balloon payment on second loan	3,614
	$33,090 Total

Disbursement

$83,200 new loan
-$33,090 consolidation
$50,110 Total

The disbursement of $50,110 is the net spendable cash based on the owner's equity. By consolidating the $83,200 at 8½% for 25 years amortization, he has that cash and monthly payments of $670. That is $292 more than he has been paying.

The owner's past payments on the 1st and 2nd trust deeds or mortgages may have been of low tax advantage. But in assuming $670 in monthly payments he can deduct $568 of that as interest, and have an annual deduction for interest of $6,816. If his taxable income is between $24,000 and $28,000, his 36% tax bracket will save him $2,540 a year or about $204.50 monthly on his income tax. In effect, his monthly mortgage expense will be reduced by that amount from $670 to

$465.50. He will now pay $87.50 more with the new mortgage than before; or, $465.50 less $378 equals $87.50.

The $50,110 cash costs only $87.50 a month, or $1,050 annually. His tax advantage is equal to an interest rate of 2.1% and is available in normal interest deduction procedures. This can be accomplished with equity built up over a period of time, the best LVR available to the owner, the best annual percentage rate he can obtain and the best bargaining position based on an area-wide sales survey. It is an excellent procedure for those with an open-end mortgage where some closing costs can be waived or the same interest rate might be applied.

Sample Lender Survey

Two separate lender surveys are demonstrated. Column 1 shows the 1977 *refinance* loan commitments based on lender policy current at the time of the loan survey, and the appraised value. This includes Lender C's annual percentage rate of 9.13%, and the $83,200 consolidation loan.

Column 2 shows the 1972 *purchase* loan commitments made when the loan prospect was the buyer of the same property. Lender B in Column 2 would have been used for the original purchase had it not been for the opportunity to assume the existing 6½% loan at the tremendous savings in interest in excess of $5,000 over Lender B's 7.8% loan during a period of, say, the following 5 years.

Shown below is the difference in savings between the two options our buyer was facing in 1972.

Option No. 1

Assumption of existing $35,000 loan @6½%, 60 payments (5 years) of $335.52 monthly, plus $4,250 2nd t.d. or mortgage @10%

	Interest	Principle	Balance
Existing loan	$ 9,947	$10,592	$24,500
2nd loan	2,125	1,272	3,614
	$12,072*	$11,228	$28,114

Downpayment	$12,750
Existing loan	35,000
2nd loan	4,250
	$52,000

*Only 7% more interest dollars than principle dollars

Option No. 2

Lender B's $40,400 loan @7.8%, 60 payments (5 years) of $320.78, with APR of 8.5% based on 5-year pay-off

Interest	Principle	Balance
$15,003**	$4,253	$36,147

$39,250 loan proceeds with 3.0% points discounted equals $40,400, or $39,250 divided by .97

Cash down payment	$12,750
Lender B's loan proceeds	39,250
	$52,000

**72% more interest dollars than principal dollars

The Loan Settlement Cost Worksheet

Below is a loan settlement cost worksheet. Use it after making a loan survey as shown above. Administrative costs are incidental to any refinance as well as to any new encumbrances placed on property involving loans. The lender's survey and the loan settlement cost sheet together are the basis for your decision to choose the best loan deal. Select your lender based on the best LVR, annual percentage rate, length of loan, monthly payments and loan fees as listed below.

A lending policy survey form is found in Appendix II.

SETTLEMENT COST WORKSHEET

Administrative services for completing a refinance loan

(Fill out for each lender)	Lender A	Lender B	Lender C	Lender D
Loan Processing Items				
1. Appraisal fee				
2. Credit report				
3. Lender's inspection fee				
4. Mortgage insurance application fee				

Items Requiring Lump Sum Payment

5. Interest from________to________@$________per day
6. Surveyor's fee (if required)____________
7. Hazard insurance premium for__________years to________

Reserves Deposited With Lender

8. Hazard insurance__________months @$________per month
9. Mortgage insurance________months @$________month.
10. City property taxes________months @$________month.
11. County property tax________months @$________month.
12. Annual assessments________months @$________month.

Title Charges
13. Abstract or title search__________ to __________
14. Title examination__________ to __________
15. Title insurance binder__________ to __________
16. Document preparation__________ to __________
17. Notary fees__________
18. Attorney's fee__________ to __________
19. Title insurance__________ to __________

Government Recording and Transfer Fee
20. Recording fees (mortgage/trust deed) $__________
21. City or county tax stamps, mtg./t.d. $__________
22. State tax/stamps, mtg./t.d. $__________

Additional Settlement Charges
23. Pest inspection report__________ to __________

---

24. *Total settlement charges for refinance*

A recent survey by the Department of Housing and Urban Development reveals an average total settlement cost of $558 based on costs of government sponsored loans. This reflects trends for conventional loans as well. Very expensive property loans can run much higher.

The definitions and explanations for settlement items below are numbered to correspond with your worksheet. Specific settlement services vary between lenders, so adapt this worksheet to the lending services available in your area.

1. *Appraisal fee* This pays for a statement of property value made by a professional independent appraiser or by a staff appraiser within the lender's organization. The lender, by this means, is assured of recovering his investment in the event you fail to repay the loan according to the contract, since he must then foreclose the loan and take title to the property. Since appraisals are considered confidential information, few lenders will volunteer to furnish a copy for you. The appraisal, however, may contain pictures and other information which could be of value to you and will contain pertinent data upon which the appraiser based his opinion of value. Ask the lender for your copy of the appraisal report or at least demand to

review the original. If you pay a separate fee for it, you should be able to inspect this document as well as other documents you have bought and paid for. From a public relations standpoint, the institution should cooperate with you if you are really interested.

2. *Credit report fee* A credit report fee covers paperwork and the time involved in obtaining a meaningful outline of your financial risk based on how you have handled other transactions. The lender analyzes this report together with information you have given him regarding your income, outstanding bills, employment, etc. From these he will determine if you are an acceptable credit risk, and uses them to help him determine how much to lend you. If it appears to you that the terms of your financing have been adversely affected as a result of a credit report, you have the right to inspect the summary of that report. There may be a small fee for this special request. The report can be challenged as to its accuracy, and you can require corrections to be made. By contacting the Federal Trade Commission in Washington, D.C., or the nearest FTC regional office, you can obtain more specific details on your credit report rights. The FTC Buyers' Guide No. 7, "Fair Credit Reporting Act," is a good summary of the act.

3. *Lender's inspection fee* This fee is usually charged in new construction projects, for an inspection made by the lender's staff or by an outside inspector. The fee is sometimes included at the time of the appraisal, or in the case of existing structures, may be eliminated as a special item of service required for the loan.

4. *Mortgage insurance application fee* If private mortgage insurance is ordered for you at your request, this fee covers the processing of the application for the policy. This is an optional item in conventional loans.

5. *Interest* At settlement time you will be required to pay the interest that accrues on the mortgage from the date of settlement to the beginning of the period covered by your first monthly payment. Let's say your settlement date is April 16, and your first payment becomes due on June 1st. To collect just interest charges for the month of May your lender will collect interest on the settlement date for the period from April

16th to May 1st. For instance, on a loan of $83,200 at 8½% from Lender C, Column 1 of the Lender Survey, the interest would be $294.67. (83,200 times .085 divided by 12 equals $589.33 for one month. A half month's interest (15 days) is $294.67, or $19.64 for 15 days equals $294.67.)

6. *Surveyor's fee* The lender or the title insurance company may require that a survey engineer be used to determine special problems arising from easements or rights of way that seem related to your property.

7. *Hazard insurance premium* Your protection as well as the lender's against loss due to fire, windstorm and natural hazards is covered in this premium prepayment. Your home owner's policy (if you have one) insures against additional risk which may include personal liability and theft. A lender may require payment of the first year's premium at settlement. If your property is in a flood-prone area it may be necessary to insure for this special coverage, as normal hazard insurance may not protect you from this loss. In some areas of the country identified by the Housing and Urban Development Agency, you may be required by federal law to carry special flood insurance on your property. Such insurance can be obtained at a low federally subsidized rate in participating communities under the National Flood Insurance Act.

8. *Hazard insurance reserve* After determining the amount of money that must be placed in the reserve in order to cover the next due date, the lender considers this item completed.

9. *Mortgage insurance* If you default on your mortgage, the lender is protected with this type of insurance. He may ask you to pay this premium on the day of settlement to cover a certain number of months in advance. You are in a particularly good bargaining position when negotiating this insurance because you can obtain a larger loan by using this insurance as a way of lowering risk. The lender may be susceptible to interest rate flexibility here.

Do not confuse mortgage insurance with mortgage life insurance, or credit life, or disability insurance coverage. Item 4 may be one of these special policies designed to pay off a mortgage due to a possible physical disability or death of the borrower. The term "credit life insurance" is used when you borrow money for a car, appliance, or for items of furnishing, and the policy would pay off the debt if you died or became disabled. Although you are not apt to encounter credit life while negotiating a refinance of real estate it is good to be forewarned anyway. Make sure you know what you are placing your signature on. Frequently, consumers actually sign up for this type of insurance, and pay it, without realizing it. Look over every paper when the lender places the "X" at a designated line.

10. and 11. *City/county property taxes* Depending on your lender, you may have to make regular monthly payments to the reserve account for property taxes. In many cases this is a convenience to the borrower, because he is saved the trouble of submitting property tax installments separately each assessment period. However, you may want to know how your lender uses this reserve tax account. Will it be placed in your "savings" which collects interest, or have you made these dollars available to him for the institution's reinvestment before taxes are due?

12. *Annual assessments* This is a reserve item covering assessments that may be imposed by subdivisions or municipalities for special improvements such as sidewalks, sewers and paving.

13, 14, and 15. *Abstract or title search, title examination and title insurance binder* These charges cover the costs for the search and examination of records of a previous ownership, transfers, vesting of ownership and the like. The lender must determine if the borrower can furnish clear title or if any adverse matters of record are disclosed that could affect the lender's investment. Title problems can include unpaid mortgages, judgements or tax liens, conveyances of mineral rights, leases and power line easements or right of ways that could limit the use and enjoyment of the property. In some areas of the country a title insurance binder is called a *commitment to insure*. In the case of refinance, a lending institution requires a lender's title policy usually written for the amount of the mortgage which would cover his loss due to defects or problems that were not identified by title search and examination. The lender's title policy, however, is a separate item of expense noted on the work sheet.

16. *Document preparation* A separate document fee may be charged covering the preparation of the final legal papers such as the mortgage, deed of trust, note, deed, and the like. Here is where you should take care to see that these services, if they are charged to you, are not also covered under some other service fees. For example, you may have ordered private mortgage insurance through the lender, and it may have been covered in the appraisal fee or in the lender's inspection fee.

17. *Notary fee* This is a fee for having a licensed person affix his or her name and notary seal to various documents, authenticating their execution. Those drawing up the documents must sign in his presence.

18. *Attorney's fee* You may be required to pay for the lender's legal services in connection with the settlement, such as an examination of the title binder or loan contract. If a lawyer's involvement is required by the lender, sometimes his fee is consolidated into the origination fees making up the annual percentage rate commitment you obtained during your lender's survey. In that case, the fee should not be stated in the work sheet.

19. *Title insurance* The total cost of the lender's title insurance policy is shown here.

20. *Recording fees* You pay the fees for legally recording the new deed and mortgage (or trust deed), and they can be fairly large. These fees are set by state and local governments.

21. *City and county tax stamps* These taxes are based on any cash made as a down payment for real estate. Here they represent a transfer of cash in the case of refinance. They are not required where a loan is assumed or a second loan taken back by a seller of property.

22. *State tax stamps* Some state and local governments have replaced the Federal Stamp Tax program which was abolished in 1968. The rate charged by the federal government was $1.10 per thousand, or 55 cents for each $500 or fraction thereof. Various taxing authorities may have instituted this same rate for additional revenue. Your lender will know the rate you must pay.

23. *Pest inspections* Lenders vary in their requirements as to pest inspection. If a pest or termite "clearance" is required, the report itself costs a small fee. Higher costs lie in the

correction phase of the report, however. If the lender insists on your complying with certain ''recommendations'' which require you to repair damage by termites or other pests, this item could have a higher cost.

Private First Trust Deeds and Mortgages

In nearly all cases of refinancing based upon equity funding institutional lenders will accomodate the owner of residential property. But sometimes it is possible to find a group of individuals whose interest in equity funding is syndicated and whose terms are competitive with the institutional lenders. Private loans are made on all types and conditions of real property. Undedicated streets, undersized lots, smaller structures and other types of odd property may interest the private investor when commercial lenders find them unsuitable. Vacant lots in suburban areas and high risk properties are the private investor's portfolio. More work is involved in finding a lender in these instances, public or private; and the search for a private lender for over, say, $40,000 is usually difficult on a first trust deed or mortgage loan.

While the advantage to a private investor is a promise of a higher yield than found in other forms of investment, mortgage investment has in the past been restricted to a 10% interest in some states. This can be compared to 7¾% interest in an insured savings account left to maturity to obtain such a yield. And savings are not periodically reinvested as in a mortgage.

The advantage to a borrower on a private first loan is the speed with which it can be obtained once your lender is found. Also, you avoid the usual lender costs, fees and points. To eliminate the middleman and avoid broker's commissions, place your money needs in the classified section of the newspaper rather than relying solely on the yellow pages of the telephone directory. Private lenders seldom advertise for a borrower; they often place their loans through a mortgage broker.

On hard-to-place loans you may have to accept shorter terms, such as seven to ten years duration. A typical private first loan might be $25,000 - $30,000 for ten years at 9½% - 10% interest. The payments would be considerably lower than an amortized loan, and a pre-payment bonus may be included in case of sale before the loan is paid off. You can expect a pre-payment bonus or penalty to be based on 180 days' interest on the remaining principle balance at the time of the repayment or sale of the property. Your rate of payment can be much more convenient than an amortized loan, but there may be a sizable balloon payment. (See Appendix IV, Rate of Pay-Off Formula.) A private first mortgage is usually based on a rate of payment equal to 1% of the original loan amount. But the percentage can be more or less depending upon your private lender's viewpoint, your property, credit, or other such factors. A rule of thumb on the LVR for a private first loan on your property is about 85% of the appraised value.

The next chapter will provide examples of low cost borrowing secured by private second loans as well as private first loans. There is no better way to explain why private investors are in competition with institutional lenders in first trust deed loans or mortgages than to say that some of us are born speculators—and the tax treatment, the informality of operations, the possibilities of early pay-off penalties (thereby increasing the effective interest rate), the periodic yield for reinvestment, or the love of money management are all appealing to the private lender. In past years the private investor or syndicator has been limited by usury laws in some states, but has had opportunities to make real estate investments where banks or savings institutions would not take the risk. Tight money conditions can afford opportunities for private mortgages if money is not available to speculators and builders, or to those who are refinancing some existing property. There has also been a large spread, usually 3% or more, between interest rates charged by private lenders and those prevailing among institutional lenders. Nowadays the spread has narrowed considerably; still, rather than impounding a lump sum in a savings certificate account where it might remain at compound interest, private mortgage investors elect a part of the principal and/or interest to be returned. This "cash flow" can be reinvested in new mortgages thereby maintaining the constant high yield. Some usury laws limiting excessive interest have been abolished for some private mortgage loan dealings. This allows supply and demand to set the rate to whatever the market will pay for private money.

The following yields compare the differences in a hypothetical lump sum investment between a current insured savings account and an equal amount invested in a first mortgage or trust deed.

1st Mortgage Loan	\$10,000 mortgage @10% interest
	10 years - - - - - - - - - - \$10,000 yield

—Interest only payments for reinvestment—

Savings Certificate	\$10,000 savings account @7¾% interest
(compounded daily)	10 years - - - - - - - - - - \$11,704 yield

—No withdrawals for reinvestment—

The savings certificate shows an advantage of only \$1,704 over a mortgage loan investment in which interest payments are reinvested to work again at 10% or possibly an even higher yield under scarce money conditions. Any investor will realize that a passbook rate yields a mere return of 5¼% interest. The funds committed to a certificate account must remain untouched to yield the 7¾% return after ten years.

3

The Hard Money Lender

Hard money is an actual loan of funds based on your equity. It is advanced in cash form and is secured by your note and trust deed or mortgage. These junior mortgages or second trust deeds, also known as junior liens, are the portfolio of the hard money lender.

A hard money loan is distinguished from an extension of credit, such as a purchase money trust deed or mortgage taken back by a seller as a part of the purchase price of the property. For example, our buyer in the 1972 transaction of $52,000 might have been faced with an "all cash" problem had the seller insisted on a full cash deal. Although a hard money second trust deed loan could have been obtained, the hard money loan in that case would be cash advanced for a purchase, or for a purchase money trust deed or mortgage. If it was *not* to be used for a purchase, it is then not considered a purchase money trust deed or mortgage, but a *hard money equity loan.*

The reason for classifying legal terms where a note represents a debt lies in the final disposition of the debt in case of default. The basic obligation of a buyer may arise in two different ways. First, he may be lent money. Second, he may be merely extended credit by a seller who takes a note and trust deed or mortgage as a part of the purchase price. Possibly there cannot be a "deficiency judgment" where there is a "purchase money" mortgage or trust deed. A purchase money mortgage or trust deed includes both the extension of credit by a seller and the lending of money, specifically for buying property. In contrast, one who actually lends money *may* get a deficiency judgment (such as in a hard money loan) upon default and foreclosure, if it was not in connection with a sale.

There is no such term as "soft" money; but if there were, a "discount" buyer, not a hard money lender would use the adjective. Again, if the discount buyer eventually had to foreclose and the trust deed or mortgage that he bought was originally an extension of credit or money that was actually loaned to complete a sale, then he too could not get a deficiency judgment in case a foreclosure sale did not cover the debt.

A list of customary instruments of the various states is found in Appendix IX, indicating methods of foreclosure procedures. Some states permit deficiency judgments and some do not.

Equity Funding and the Junior Lien

There are many situations in which equity funding is called for. Sometimes you must raise money when very little equity remains in the property, and complete refinancing is either unavailable or too expensive. Speedy procedures, simplicity, the need for a temporary solution to a cash problem and flexibility of negotiations are typical reasons for borrowing funds secured by junior liens.

A mortgage broker, a bank, or a savings and loan association is not necessary here, and in fact may raise your costs above those of private lenders.

Banks or savings institutions may not charge you any points or one-time costs in their junior lien procedures or equity funding transactions. But the interest rate is comparatively high. The payments are usually amortized on a simple interest basis, based on 12 or 15 year term. The annual percentage rate is high under such circumstances.

If you use the services of a mortgage loan company, there is a commission on the amount of principal borrowed, usually

10% or 15%. This is separate from the interest rate. However, all settlement costs and commissions are incorporated into the principal amount of a loan broker's negotiated amount to be borrowed. This loan is obtained from private individual junior lien investors who have made the money available to brokers to invest for them. You can do this yourself. It is less expensive to turn directly to private investors.

An Example of Equity Funding

Here is an example of equity funding and the role of the hard money lender. Let us assume that our buyer in 1972 does not choose to take advantage of the existing $35,000 loan in spite of its attractive 6½% interest and the lender's approval. (See Chapter 2.) He has to consider the junior lien and the balloon payment due in five years. And perhaps the seller would offer the salable second loan on the market anyway at a 25% discount. Why not apply for Lender B's commitment of $40,000 in the Lender Survey and combine the available down payment up to a discounted price of the home—which is equivalent to what the second trust deed or mortgage would have sold for if discounted? After all, the buyer reasons, if this is to be a cash sale he will ask the seller to subtract the discount equivalent of $1,062.50 from the $52,000 price. Since the $4,250 second loan would be sold anyway for 25% off (or $1,062.50 less than its face value), why not offer all cash to the seller by an amount of $1,062.50 less than the $52,000? The deal was made on this basis and title was passed for $50,937.50 with all parties happy with the transaction. The buyer's equity is the down payment of $11,687.50.

The new buyer's problem, though temporary, is that the property needs one more bedroom. Plans to remodel can be put off, and ordinarily should be, while equity builds up. The owner's income is adequate but the bank account is too low for a withdrawal so soon after purchase of the property. But in his enthusiasm for the idea the new owner decides to check recent sale prices of homes in the area and uses a market sales survey form like that found in Appendix I of this book. From this he concludes that the $50,900 was the property's highest limit of value at that time. He sees that, at most, the $11,700 down payment is the only recognized equity.

But three years later, in 1975, the owner's new sales survey reveals a 19.35% annual increase in property values in his area and the equity of his property has thereby increased to over $29,500. With the added $1,200 paid on the mortgage and the original $11,700 down payment, his equity now is around $42,400. The owner's figures are as follows:

$80,400	indicated 1975 value
-$42,400	equity
$38,000	principle balance remaining on loan

At one time it was considered a fearful thing to obtain a second mortgage or trust deed, and bankers were the first to tell you so. Now, the business is good for bankers because it is a moneymaker. Banks were not willing to risk a loan of any kind of property of doubtful resale, so private investors had little competition in the field of junior liens. Banks now will invest in second mortgages up to an amount totalling 80% of the appraised value. This includes the amount remaining on an existing first loan. Example: A property worth $80,400 with an existing first mortgage of $38,000 against it will justify a second mortgage of $26,320. The total of the two loans would be $64,320; or 80% of its value. This is not a purchase money loan. It is a bank's straight hard money loan commitment with no sale involved based on a top encumbrance of 80% of value. The value of the property must be verified by a separate appraisal before this hard money equity-funded bank loan can be made.

In contrast, the private investor's straight hard money loan commitment where no sale is involved uses 50% of the equity as a limit to cash advances. A top loan for the above owner would be $21,200 based on one-half of his $42,400 equity. This rule of thumb for private investors became customary over the years before banks became competitors in this second mortgage investment field. The appraised value is always the factor on which the percentage is calculated in straight hard money loans.

It is no wonder that banks and savings institutions are stiff competitors on junior liens. But the private lender does not demand loans with an annual percentage rate including fees.

And while some states restrict private lenders to a 10% interest limitation, banks can increase the annual percentage rate by increasing the interest rate.

On the other hand, if a sale is conducted allowing a buyer to assume a low interest first mortgage loan of $38,000 remaining on the property, a tremendous gap with an extreme down payment would then be required on the $80,400 sale. To fill part of that gap the buyer has only $16,080 or his 20% down payment. Since the particular deal is a purchase, not simply a loan on equity, the banker or commercial lender will offer a high-interest new first loan on long amortization rather than advise you to take over or "assume" the existing low interest first mortgage. But a private lender or hard money lender will limit his risk on this sale by offering a cash advance in an equal amount to the buyer's down payment - another $16,080. The private lender in this transaction would be making a cash loan secured by a purchase money second trust deed or mortgage. It is a hard money loan used for a purchase. This private lender is depending upon the buyer's integrity and the actual sale price to determine the value of the property. In this case there still remains a $10,240 difference, since the hard money lender loans no more than shown by the down payment. Now the question arises whether or not the "seller" is willing to carry back a purchase money third trust deed or mortgage as a credit advance. This extreme example is for illustration only, but it shows the difference between a purchase money loan and the straight hard money loan where no sale is involved.

How to Advertise for a Hard Money Equity Loan

If you need a hard money second trust deed loan in an amount no higher than ½ your apparent equity, a simple want ad in the local newspaper will produce private investors. In nearly all states the customary instrument can be advertised as a trust deed or mortgage. See Appendix IX for specific listings of the types of instruments used in each state.

You can make final legal inquiries when you meet with the lender in escrow or in the lawyer's office. Your questions concerning title ownership, default penalties or settlement costs should be thoroughly answered before you sign the documents. Abstracts of title drawn up by an attorney, rather than a title insurance policy, must be furnished in some states before a loan on your property can be recorded. A clear title protects your investor so that his investment is recoverable under the law in case of default.

You must convince your lender of the value of your property, your good credit and general integrity. These ads are some examples of good salesmanship:

Advertisement	*Comment*
Money Wanted. - Real Est. $14,500 2nd TD 10% 5 yrs. on $80,000 home Pr/pty 000-0000	This ad indicates plenty of equity with definite terms.
$20,000 10% pay $200. mo. 50% equity, pre-pay bonus or int. only Princ. only P/pty 000-0000 After 5	Suggests a 1% rate of payoff with terms negotiable. Interest only (without due date) makes phone ring.
or	
$21,200 2nd mortg. secured by $80,400 value due 1 year builder's 10% 000-0000	A builder's slow mover will attract investors for short periods.
or	
$12,000 2nd TD 10% due 1 yr. or amortize. Westside area Pr/pty 000-0000	A strong indication to suggest special terms.

Here are some further suggestions on advertising for a hard money second trust deed or mortgage loan.

1. Leave no doubt in the reader's mind that you seek a private investor only. "P/pty," "Princ. only," "No brokers please" are the most popular words.
2. Emphasize value with your own dollar estimate or with the location of the property.
3. Be prepared to discuss your opinion of equity if a dollar amount is in the ad.
4. Make sure your ad is placed under the correct classified

heading. An equity loan for no more than one half of your equity belongs under "Money Wanted" or "Money Wanted Real Estate." You are selling a loan to hard money lenders that has not yet been made.

5. Never try to obtain a junior lien for less than par from private investors through ads. You can negotiate a lower interest rate in person later. Securities are at par when market price equals the par value.

6. When you meet the investor you must convince him not only of the property's value but of your good credit and general integrity.

7. Do not insist on your own escrow company, lawyer or title company. Make the agreement on these particulars a mutual one.

4

Discounts And Hypothecation Loans

An excellent example of a small investor's dream is the proposed second trust deed our seller in Chapter 2 was planning to put on the market in 1972 while the $52,000 sale of his property was about to be processed. The seller of the property is the key to the price of junior liens, whether the times indicate a seller's market or not. The market in junior liens often varies from 10% to 50% off the face value, and the investor will try to make the most of the discount. The 1972 seller, realizing that the buyer of his property had a limited amount of cash, set a limit of no more than 25% off on any second loan he may have to take back. So he put an ad in the classified section of his paper under "Mortgages-TD's for Sale."

18% YIELD
$4,250 2nd TD due 5 yrs.
10% int. 25% disc. on
$52,000 home. p/p 000-0000

Through this ad an amount of $3,187.50 could buy the trust deed with a balloon payment due in 5 years with an 18% yield overall. See Appendix V.

It was offered to an investor who is a *discounter* rather than a hard money lender, although in many cases they are one and the same. Discount operations, although more profitable, are not as numerous since most are connected with a sale and their investments are considered purchase money mortgages. Since the rate of payment on a purchase money second mortgage is usually 1% or more per month including interest, the buyer of the property can pay the 1% installment rate or he can pay the complete unpaid balance anytime without a prepayment penalty. In most cases there is a balloon payment. If the face amount of a purchase money mortgage or trust deed taken back by the seller is equal to the down payment made by the buyer, a discount can be had for between 18% and 25% off. With very low down payments reflecting an excessive second mortgage, the discount can run as high as 50% off or more.

Seasoning the Junior Lien

A note is *seasoned* when it is held long enough to prove its value. A discounter usually considers a seasoned junior lien of more value to him, and will pay more for one that has a consistent payment record. This appears to him as an investment with less risk which is therefore worth a smaller discount. Many builders prefer to cash out the notes at discount as they market their projects, thereby supplementing their operating cash in order to begin other building projects. But you would do well to season these second mortgages. Frequently builders, contractors, and developers take back second loans which are purchase money mortgages or trust deeds. They soon sell them to discounters or sell them almost immediately after the project is sold. If they did not need a turnover of capital or operating cash, they could simply retain the mortgages for added income. As the notes season, they become more valuable to an investor; the installments will probably continue to be paid as they have in the past. Therefore, investors will pay more to purchase these notes for, say, 15% discount rather than earlier at 20% or 25% of their face value. As regular payments are made the face value is lowered because the principal balance of the notes is being paid down. In the meantime the builder is collecting payments plus the interest carried by the junior liens. As another approach to lowering risk, an experienced private discounter would rather buy many small junior liens than a few representing large individual notes.

The seller of the note must know how sensitive a discounted mortgage or trust deed can be. For instance, when this junior lien is sold in escrow to an investor there is no evidence that the amount received from the sale of the note actually represents its potential worth. In the transferring period before escrow actually closes there can be more than one sale of the note; it could change hands with a new and smaller discount each time before the final buyer settles down to collect the payments from the buyer of the property. This is legal and quite common. If sellers of real property can wait for the buyers to pay on the second mortgage for a substantial length of time before "farming it out" to a discounter, the note will bring a higher price because it has been seasoned. But if the seller of the property needs money as soon as possible, he would rather sell or "discount" the note while the property transaction is undergoing the escrow process.

The second trust deed or mortgage is a supplementary income when the note is held by a seller of property. The principal and interest payments are submitted periodically during a time period agreed to between buyer and seller, just as a buyer makes installments on his first mortgage. In the long run—during the period of payment the interest has accrued—the seller will have obtained more money for his property than if he had sold for cash. He is also seasoning the second mortgage for any future discounter in the event he later needs to raise cash. Or he may be interested in *hypothecating the note* at a later date.

Hypothecation Loans

Hypothecation means giving something as security without actually giving up possession of it. Rather than selling the note, he can borrow up to 70% of its remaining face value and still retain its ownership and income.

You do not need a bank or a mortgage loan broker to obtain a loan on an interest bearing second mortgage. Your ownership of the note is collateral just like any other property you possess. The investor who advances money on your collateral note is called a hypothecation lender. He now collects periodic payments from you after advancing a lump sum to you as a loan. A hypothecation lender can be attracted with an ad under "Money Wanted" in the classified section of a newspaper. As an investor, he seeks a higher interest on his money. Terms can be negotiated as "interest only" payments, or the rate of pay-off convenient to you and a balloon payment at the end of the term just as in all hard money loans of short maturity. All hard money lenders, including collateral lenders, nearly always insist on a prepayment penalty. The loan you are seeking should sell as in the following illustration. Here, the purchase money second mortgage has been paid down to a face value of $2,800. Installments have been regularly made for the past two years on an original 5-year purchase money junior lien at 10% interest. The original face value was $4,250.

Money Wanted - Real Est.
$2,800 seasoned collateral loan
10% int. w/pre-pay bonus P/pty 000-0000

The pre-pay bonus is the key here. The investor will determine the amount to offer you based on the original amount of the second mortgage. The customary maximum is 70% of the remaining principle balance or $1,960.

The title officer or escrow service will use the standard *collateral security note* in assigning the existing note and trust deed or mortgage to the lender. The assignment of the note can revert to you on full payment of the loan. In other words, the collateral will be used as a lever. If you are the holder of a junior lien and wish to keep the security rather than sell it at a discount, a collateral loan is for you. This is also the case if you do not need the full amount that the junior lien would bring on a discounted sale, but a lesser amount. For a hypothecation lender to consider your junior lien, it must be well seasoned for some time after you sold the property that secures it. The payments continue to be paid to you, the record holder of the junior lien, and not to the recorded hypothecation lender. Although the terms are always negotiable, most collateral loans call for a prepayment clause similar to that in a first loan on real property. There is no reconveyance deed at the time of pay-off because the deed is still in your name as a matter of record. Although there is an assignment of the junior lien, it does not convey ownership but is held simply as collateral for the new loan. This is an example of leverage.

Leverage is a word used when financing or refinancing in order to buy something or invest in something with somebody else's money. Any construction project for resale at a hoped-for profit which uses another's capital in order to take advantage of that business opportunity is leveraged. In the above case, the new collateral loan can be paid off after its use while you still retain the original purchase money trust deed or mortgage and continue to receive the original installments from the buyer of the property you sold.

5

Equity Loan Procedure

The lender that responds most favorably to your lender survey on refinancing asks you to drop by at your convenience. By now he is fairly sure why you want to loosen up your equity: to obtain cash for a building project. But as far as the lender is concerned the refinance is based solely on the real estate, the mortgagor's property value, and not on any investment to be made from the funds advanced. Practically speaking, the lender is risking his money on your basic integrity and the present security of your existing property. What is done with the funds interests the lender only as far as it reflects greater or less risk that they will be repaid. In the case of an applicant's plans to later purchase a building site with all or part of the money, since there are no contractual agreements between the borrower and the lender in this matter, the lender can only judge the risk in terms of granting the refinance loan. You, the applicant, may want to impress upon the lender the stability of your building plans—and he will probably consider future business with you in a construction loan later on—but these building plans play no part in the refinance now being negotiated.

Understanding the lending institution's internal loan-processing system will take some of the mystery out of applying. Large lending firms follow what is called the *unit system.* A department head takes the application, analyzes it, and coordinates the loan processing. He then sends it to the loan committee which can be a single person or a small group who handle specific details necessary before the application is presented to the loan closer. At this point a determination is made as to whether the lender has a good first mortgage loan for its portfolio. Other savings and loans may use a different system. They feel that since the initial contact is a vital link between the lender and the prospective borrower, a loan officer must interview all prospective borrowers. Still other lenders permit a receptionist to conduct the first interview and a qualified loan officer to conduct the subsequent interview.

The loan servicing department is concerned with payment problems on an ongoing basis. Some lenders instruct their interviewers to point out that, in the event of any inadvertant collection problems, there are some alternatives. That is, the interviewer may say that if the borrower has met all the payments on the new loan promptly and needs funds for remodeling projects or for any unforeseen problems in a new venture, the lender can again refinance the loan to the original amount at very little cost. The lender will agree to this provided that the property continues to be maintained as well as in the past and its value continues to hold. In this event, the loan servicing department will relax its efforts to stimulate collection from the borrower. But the interview period is not the time for the loan officer to make threats about the importance of making payments, and he knows it. A good interviewer is discrete in his explanation of collection and servicing problems.

At the time of the interview the loan officer, or whoever is assigned to you for the purpose of taking your application, goes over the estimated costs of the refinance. He should furnish you with the specific details of your loan at this point—the rate of interest, the APR, the escrow administration costs, title search and other costs including an appraisal and credit report fee. Remember that all fees and costs are not included in the loan you must pay back, and if you go ahead with the application you may be asked to advance the appraisal fee and the credit report fee. You should be told at this time approximately when your loan will be approved, the date of closing, the amount of your payments and when they will be due. If all these points are covered in the interview, time-consuming call backs can be cut down. Spending more time in the interview saves you time later while clerical people run down your application in answering your inquiries.

The Preliminary Application

Many lenders receive information from the borrower on a preliminary application, screening out at a glance those who would be automatically turned down if they were to make a formal loan application. The lender analyzes it to see if you meet the qualification to fill out the formal application. The preliminary application includes your name and address, the address of the subject property, the purpose of the loan and your estimate of the property's value. This is the first place to use your independent sales survey and your equity estimation. It also includes personal data such as income, age and financial obligation, as well as estimates provided by the interviewer of such items as monthly loan payments, loan charges and fees.

Lenders' Promotional Tactics

Later in this chapter various types of lenders will be discussed. But promotional tactics of lenders are worth knowing before you fill out the final loan application. In promotional letterheads, an institution states whether it is a state or federally chartered lender. These letterheads usually carry in small print the assets, the liabilities and the name of the government agency controlling or insuring their operation. The various lenders or their exclusive agents should always have their operational status fully exposed on the letterhead. A general agent will not have an exclusive lending institution on his letterhead unless he is, in fact, an exclusive correspondent.

Generally, a mortgage loan broker promotes his services by advertising particular types of loans which may have a lesser rate of interest, therefore offsetting the fee to the borrower. He is a finder of loans that the borrower may not have the time, energy or source knowledge to locate for himself. The person who negotiates a loan on real estate for a buyer, for compensation, is a general real estate broker. But in some states the real estate broker who also negotiates loans that are not incidental to the ordinary course of a real estate sale or exchange in fact is called a mortgage loan broker. If the broker operates as a middleman for the public's advantage, determine precisely the extent and kind of his services—is he an independent broker or an exclusive agent, correspondent, or insurance company representative? Examine his card and letterhead, in addition to other verification. They should not be vague. You should determine his capacity as an employee or an independent. Some commercial banks make what are considered high-risk loans through a preferred list of loan brokers and charge whatever the market will bear in points. In any case, the final settlement sheet indicates full disclosure as to who gets what and for what services.

All the information on the preliminary application will be obtained from you to be developed and interpreted by the interviewer. Since no credit report or verification of your financial condition is needed to complete this application, you will be informed on your initial visit whether or not a formal application can be considered. If the interviewer clears you, he may suggest that you fill out the formal application immediately.

The Loan Application

The final application is more formal and asks for more detailed information on you and the property. A complete legal description of the property must be furnished, including a description of any existing buildings and any mortgages or liens on that property. If you are not familiar with a legal description, it can be found on the property tax form or on the title policy you bring to the lender. Your personal credit statement, previous employment, present residence, credit references, life insurance and other assets, and the purpose of the loan are required information on the application. In this case, state the purpose as refinancing an old loan.

The formal application does not mean that you have to go through with the loan regardless of whether or not you wish to

RESIDENTIAL LOAN APPLICATION YOU MAY APPLY SEPARATELY OR JOINTLY WITH YOUR SPOUSE

MORTGAGE APPLIED FOR | Type ☐Conv. ☐FHA ☐VA | Amount $ | Interest Rate % | No. of Months | Monthly Payment Principal & Interest $ | Escrow/Impounds (to be collected monthly) ☐Taxes ☐Hazard Ins. ☐Mtg. Ins. ☐

Prepayment Option

SUBJECT PROPERTY

Property Street Address | City | County | State | Zip | No. Units

Legal Description (Attach description if necessary) | Property is: ☐Fee ☐Leasehold ☐Condo ☐PUD ☐DeMinimis PUD | Year Built

Purpose of Loan: ☐Purchase ☐Construction-Perm. ☐Construction ☐Refinance ☐Other (Explain)

Complete this line if Construction-Perm. or Construction Loan | Lot Value Data Year Acquired | Original Cost $ | Present Value (a) $ | Cost of Imps. (b) $ | Total (a+b) $ | ENTER TOTAL AS PURCHASE PRICE IN DETAILS OF PURCHASE

Complete this line if a Refinance Loan | Purpose of Refinance | Describe Improvement [ ] made [ ] to be made

Year Acquired | Original Cost $ | Amt. Existing Liens $ | Cost $

Title Will Be Held In What Name(s) | Manner In Which Title Will Be Held

Source of Down Payment and Settlement Charges

This application is designed to be completed by the borrower(s) with the lender's assistance. The Co-Borrower Section and all other Co-Borrower questions must be completed and the appropriate box(es) checked if ☐ another person will be jointly obligated with the Borrower on the loan, or ☐ the Borrower is relying on income from alimony, child support or separate maintenance or on the income or assets of another person as a basis for repayment of the loan, or ☐ the Borrower is married and resides, or the property is located, in a community property state.

BORROWER	CO-BORROWER*
Name / Age / School Yrs	Name / Age / School Yrs
Present Address / Years at present address / ☐ Own ☐ Rent	Present Address / Years at present address / ☐ Own ☐ Rent
Street	Street
City/State/Zip	City/State/Zip
Former address if less than 2 years at present address	Former address if less than 2 years at present address
Street	Street
City/State/Zip	City/State/Zip
Years at former address / ☐ Own ☐ Rent	Years at former address / ☐ Own ☐ Rent
Marital Status ☐Married ☐Separated ☐Unmarried (incl. single, divorced, widowed) / DEPENDENTS OTHER THAN LISTED BY CO-BORROWER NO. AGES	Marital Status ☐Married ☐Separated ☐Unmarried (incl. single, divorced, widowed) / DEPENDENTS OTHER THAN LISTED BY CO-BORROWER NO. AGES
Name and Address of Employer / Years employed in this line of work or profession? ____ years / Years on this job / ☐Self Employed*	Name and Address of Employer / Years employed in this line of work or profession? ____ years / Years on this job / ☐Self Employed*
Position/Title / Type of Business	Position/Title / Type of Business
Social Security Number*** / Home Phone / Business Phone	Social Security Number*** / Home Phone / Business Phone

Federal National Mortgage Association Form
Figure 5-1

reconsider the decision. But your signature formally indicates that you will pay all necessary charges or processing fees on the application whether or not the loan is granted. This includes the cost of a title search, even though you have your old title insurance policy. And the signature shows that you can furnish your credit references and federal tax returns. See Figures 5-1, 5-2, 5-3, and 5-4, which show Federal National Mortgage Association Form 2003 Rev. 3/77.

The Applicant as a Credit Risk

Your past credit record can be determined through an analysis of the credit bureau report and information gained from any source from which you have borrowed. If you have

GROSS MONTHLY INCOME				MONTHLY HOUSING EXPENSE			DETAILS OF PURCHASE	
Item	Borrower	Co Borrower	Total		PREVIOUS	PROPOSED		
				Rent	$			
Base Empl. Income	$	$	$	First Mortgage (P&I)		$	a. Purchase Price	$
Overtime				Other Financing (P&I)			b. Total Closing Costs(Est)	
Bonuses				Hazard Insurance			c. Prepaid Escrows(Est)	
Commissions				Real Estate Taxes			d. Total (a+b+c)	$
Dividends/Interest				Mortgage Insurance			e. Amount This Mortage	( )
Net Rental Income				Homeowner Assn. Dues			f. Other Financing	( )
Other † (Before completing, see notice under Describe Other Income below)				Other			g. Present Equity in Lot	( )
				Total Monthly Pmt	$	$	h. Amount of Cash Deposit	( )
				Utilities			i. Closing Costs Paid by Seller	( )
Total	$	$	$	Total	$	$	j. Cash Reqd. For Closing(Est)	$

DESCRIBE OTHER INCOME

B—Borrower C—Co-Borrower	NOTICE † Alimony, child support, or separate maintenance income need not be revealed if the Borrower or Co-Borrower does not choose to have it considered as a basis for repaying this loan.	Monthly Amount
		$

IF EMPLOYED IN CURRENT POSITION FOR LESS THAN TWO YEARS COMPLETE THE FOLLOWING

B/C	Previous Employer/School	City/State	Type of Business	Position/Title	Dates From/To	Monthly Income
						$

THESE QUESTIONS APPLY TO BOTH BORROWER AND CO BORROWER

If a "yes" answer is given to a question in this column, explain on an attached sheet.	Borrower Yes or No	Co-Borrower Yes or No		Borrower Yes or No	Co-Borrower Yes or No
Have you any outstanding judgments? In the last 14 years, have you been declared bankrupt?			Do you have health and accident insurance?		
Have you had property foreclosed upon or given title or deed in lieu thereof?			Do you have major medical coverage?		
Are you a co-maker or endorser on a note?			Do you intend to occupy this property?		
Are you a party in a law suit?			Will this property be your primary residence?		
Are you obligated to pay alimony, child support, or separate maintenance?			Have you previously owned a home?		
Is any part of the down payment borrowed?			Sales Price of previously owned home?	$	$

* FHLMC requires self employed to furnish signed copies of one or more most recent Federal Tax Returns or audited Profit and Loss Statements. FNMA requires business credit report, signed Federal Income Tax returns for last two years, and if available, audited P/L plus balance sheet for same period.

** All Present Monthly Housing Expenses of Borrower and Co-Borrower should be listed on a combined basis.

*** Neither FHLMC nor FNMA require this information.

FHLMC 65 Rev. 3/77
91-6148 (Rev. 4-77)

FNMA 1003 Rev. 3/77

Federal National Mortgage Association Form
Figure 5-2

shown concern in the past for your financial obligations you are likely to continue to do so in the future. Your total obligations will be considered. It is possible that the proposed installments and your income are out of balance due to other fixed expenses such as income taxes or life insurance policies.

Next to the application for credit, the investigation and analysis of the mortgagor is most exhaustive. You will not personally take part in this follow-through, but by providing reliable information you may satisfy the lender on an incidental item.

The methods used in determining a good loan depend upon the efficiency of the credit rating procedure. But there is little or no security in a credit report. In any secured loan, the lender has the goods to sell in order to recover the money loaned in

This Statement and any applicable supporting schedules may be completed jointly by both married and unmarried co-borrowers if their assets and liabilities are sufficiently joined so that the Statement can be meaningfully and fairly presented on a combined basis; otherwise separate Statements and Schedules are required (FHLMC 65A/FNMA 1003A). If the co-borrower section was completed about spouse, complete this statement and supporting schedules about spouse also.

☐ Completed Jointly ☐ Not Completed Jointly

ASSETS		LIABILITIES AND PLEDGED ASSETS			
		Indicate by (*) those liabilities or pledged assets which will be satisfied upon sale of real estate owned or upon refinancing of subject property.			
Description	Cash or Market Value	Creditors' Name, Address and Account Number	Acct. Name, if Not Borrower's	Mo. Pmt. and Mos. left to pay	Unpaid Balance
Cash Deposit Toward Purchase Held By	$	Installment Debt (include "revolving" charge accounts)		$ Pmt./Mos. /	$
				/	
Checking and Savings Accounts (Show Names of Institutions/Acct. Nos.)				/	
				/	
Stocks and Bonds (No./Description)				/	
				/	
				/	
Life Insurance Net Cash Value Face Amount $		Automobile Loan		/	
SUBTOTAL LIQUID ASSETS	$				
Real Estate Owned (Enter Market Value from Schedule of Real Estate Owned)		Real Estate Loans (Itemize and Identify Lender)		X	
Vested Interest in Retirement Fund					
Net Worth of Business Owned (ATTACH FINANCIAL STATEMENT)					
Automobiles (Make and Year)		Other Debt Including Stock Pledges (Itemize)		/	X
Furniture and personal Property		Alimony, Child Support and Separate Maintenance Payments Owed To			X
Other Assets (Itemize)				/	
		TOTAL MONTHLY PAYMENTS		$	X
TOTAL ASSETS	A $	NET WORTH (A – B) $		TOTAL LIABILITIES	B. $

Federal National Mortgage Association Form
Figure 5-3

case of default. But no lender anticipates with satisfaction the day he may have to recover your property to get his investment returned to him.

Aside from the property as security, the lender must keep in mind how much trouble he may encounter with you as time goes on. Are you likely to be able to meet your obligations? Is there a good chance that you will be able to meet due dates? Will you have difficulty meeting payments because they become a burden to you along with your other obligations? The lender must analyze your credit standing. He may refer to his credit agency, or to a credit bureau dealing in local retail credit. He may also contact your friends along with the business people you list as references. A credit agency used as a reference may be an alternative to listing friends; if you do so

SCHEDULE OF REAL ESTATE OWNED (If Additional Properties Owned Attach Separate Schedule)

Address of Property (Indicate S if Sold, PS if Pending Sale or R if Rental being held for income)	▽	Type of Property	Present Market Value	Amount of Mortgages & Liens	Gross Rental Income	Mortgage Payments	Taxes, Ins. Maintenance and Misc.	Net Rental Income
			$	$	$	$	$	$
		TOTALS →	$	$	$	$	$	$

LIST PREVIOUS CREDIT REFERENCES

▽ B—Borrower C—Co-Borrower	Creditor's Name and Address	Account Number	Purpose	Highest Balance	Date Paid
				$	

List any additional names under which credit has previously been received ______

AGREEMENT: The undersigned applies for the loan indicated in this application to be secured by a first mortage or deed of trust on the property described herein, and represents that the property will not be used for any illegal or restricted purpose, and that all statements made in this application are true and are made for the purpose of obtaining the loan. Verification may be obtained from any source named in this application. The original or a copy of this application will be retained by the lender, even if the loan is not granted.

I/we understand that periodically you may receive information and answer questions and requests from others, like credit reporting agencies, about me/us and my/our transactions with you.

I/we fully understand that it is a federal crime punishable by fine or imprisonment, or both, to knowingly make any false statements concerning any of the above facts as applicable under the provisions of Title 18, United States Code Section 1014.

______ Date ______ ______ Date ______
Borrower's Signature Co-Borrower's Signature

VOLUNTARY INFORMATION FOR GOVERNMENT MONITORING PURPOSES

If this loan is for purchase or construction of a home, the following information is requested by the Federal Government to monitor this lender's compliance with Equal Credit Opportunity and Fair Housing Laws. The law provides that a lender may neither discriminate on the basis of this information nor on whether or not it is furnished. Furnishing this information is optional. If you do not wish to furnish the following information, please initial below.

BORROWER: I do not wish to furnish this information (initials) ______
RACE/ NATIONAL ORIGIN ☐ American Indian, Alaskan Native ☐ Asian, Pacific Islander ☐ Black ☐ Hispanic ☐ White ☐ Other (specify) ______
SEX ☐ Female ☐ Male

CO-BORROWER: I do not wish to furnish this information (initials) ______
RACE/ NATIONAL ORIGIN ☐ American Indian, Alaskan Native ☐ Asian, Pacific Islander ☐ Black ☐ Hispanic ☐ White ☐ Other (specify) ______
SEX ☐ Female ☐ Male

FOR LENDER'S USE ONLY

(FNMA REQUIREMENT ONLY) This Application was taken by ______ (Interviewer), a full time employee of ______ in a face to face interview with the prospective borrower

FHLMC 65 Rev. 3/77 REVERSE FNMA 1003 Rev. 3/77

Federal National Mortgage Association Form
Figure 5-4

and if the credit bureau or agency has a current and dependable information report available, there will be no need for the lenders to talk to your friends. Make sure the friends you do use as reference are familiar with your financial affairs.

If you experience a denial of a loan that is clearly the result of your credit standing, the denial may be simply a matter of current policy of that particular lender. The denial is strictly confidential between you and the representative who took your application, and is not in any way to be interpreted as a standard for another application elsewhere. Another credit procedure at another bank or savings association may be interpreted entirely differently and the application may encounter no difficulty at all. Often the denial is a temporary condition, and the applicant has a completely different experience with the same lender at a later date using the same information previously submitted.

If you feel that an unfair report has been submitted to the lender, there are further details for redressing your grievance in Chapter 2, "Credit Report Fee." The Fair Credit Reporting Act does not give you the right to inspect or physically handle your report at the credit reporting agency, nor to receive an exact copy of the report. But you are entitled to a summary of the report showing the nature and sources of the information it contains.

Builders are asked to give information regarding their union standing, and must usually furnish W-2 forms for the previous year or two. If you are self-employed, you may have to furnish a profit and loss statement. If you have recently established your own business, the lender may ask for some proof of your experience in that business. Salesmen should be prepared to supply W-2 forms or an income tax statement if you work exclusively or substantially on a commission basis.

The cost of a credit report varies depending on the locality. The cost will be higher if the credit bureau or agency extends its inquiry into other communities or states.

Lending Conditions

Lending institutions are flexible in some areas and inflexible in others when it comes to granting loans. They operate under fixed statutes, while at the same time they use the services of appraisers—whose methods are not standard. At the same time, lending institutions must do their work according to the supply of money available to them, which varies. If a lender can correlate its money supply with the government statutes under which it operates, its own lending policies, and the appraiser's estimate of value, it can grant a loan up to 90% of the appraised value, where a sale is involved.

No two appraisers ever reach the same figure on the same property. Appraising is an inexact science because it is not universal in its methods. Appraisers are humans who have opinions. Since our federal and state governments don't make the loan, the lending rules depend upon the appraised value made by the appraiser hired by the lender. Therefore, flexibility is built into each lender's procedures.

State chartered banks and savings and loans have their own statutes. Federally chartered banks or savings and loans also follow statutes which limit their activities. In both cases, statutory limits are set on loan to value ratios, asset limitations, the radius of the lending area, maximum term in years, interest payment intervals, principle payments, and special authorized procedures for directors of the organization. Lending institutions may vary considerably from one another in the loans they grant, depending upon the money supply and their application of the statutes.

The Lender's Appraisal

The combined information received from the mortgage credit analysis and the appraisal provides the basis for the lender's decision as to whether or not to grant the loan. This data also helps him determine the amount of the loan, the interest rate and the maturity. The appraisal is separate from the underwriting in that the appraiser is estimating a value in document form and showing the means whereby he reached that value; whereas the underwriting relates to an analysis of the borrower's credit risk, or his ability to fulfill all of the mortgage loan conditions. A lender is much more concerned about underwriting the mortgage and evaluating the property in mortgage refinance loans during tight money conditions.

There is no better determination of the value of real estate than its sale, which indicates the activity of the marketplace at the time of sale. That is the basis for your independent survey. The value of your property can be compared in the absence of a sale to that of other property like it; the lender's appraisal, though, is more complicated. Since you are not selling, and therefore there is no established sale price, the lender must satisfy himself of the value of your property before he agrees to loan you any money based on your equity. He is interested in the extent to which the property provides sufficient security in the event of default by you during the life of the loan.

The appraiser must judge the influence of the surrounding area on your property. He looks to see if the income in the community is stable and makes a forecast on the prospect of continued stability in the years ahead. When income is lowered, the ability of the owner to keep up maintenance on the property is affected. If this lasts for some time, it can affect mortgage payments and property values. The stability of the area's income is important. The appraiser looks at the industrial, business, and service occupation climate. In a one-industry town, there is a danger of collapse or relocation, or simply a lapse in production, meaning unemployment, possible defaults and a general inability to keep up property maintenance. In an area with many industries, businesses and service occupations, not all of these will be thriving at once; some property owners and lenders can expect losses. The lender takes into account the rate of business failure and the types of business which succeed and fail in examining the trends of value for the future. The appraiser may establish in his own mind the degree of pride of ownership in the community. It is important to him to know the approximate percentage of home owners to renters as well as the schools and churches serving the area. Physical and natural surroundings will be noted for attractiveness. The availability of shopping and transportation is also high on the list for analysis. The work availability and the average income for the neighborhood may indicate to him a general lifestyle of the homeowners. This does not mean that your individual habits are of particular interest to the appraiser. He is not concerned about your personal life except as it may affect the market value of the property. And he doesn't compile all his information about the area in one appraisal assignment. The data is most likely catalogued in the office by general area, or the new data is constantly collected from staff field appraisers.

The appraiser views your property just as a prospective buyer would. The property should have qualities that serve the purposes of its original intended use. The site itself is studied extensively to determine the services it has to offer the building on it. Stable values will be found in neighborhoods with similar general design—that is, where quality of construction is nearly the same and the exteriors, floor plans and equipment are more or less uniform. A home that is well constructed is generally obvious to an appraiser familiar with quality construction materials. He should know the construction difficulties in the year of the building start. Starts during periods when quality materials are scarce, such as during wartime, produce buildings that are often not as well constructed as building starts in better times. The appraiser considers in his estimate the higher maintenance necessary on such buildings.

An appraiser is more handicapped in estimating the quality of the structure after finished construction than during the framing and "wrapping" of a new project. But the beams in the attic or the joists in the basement area may indicate lumber quality. Seepage of water to the subfloors will be noted. Drainage at the downspouts can be a source of seepage trouble. Or the final grading around the structure may be too close to the building with no splash block to drain the water away from the foundation wall. The subfloor construction may show signs of seepage, indicating the lack of dampproof procedures when the subflooring was installed. The appraiser will note the strength of the support of the roof load and will carefully inspect the insulation.

He will check areas between the foundation walls and the sill to see that no openings exist. Any openings should be filled with a cement mixture or a caulking compound, and very possibly this can be a provision for obtaining the loan. This filling can prevent heat loss and can keep insects from entering the basement. Concrete block or other masonry walls exposed above grade often show dampness on the interior after a prolonged rainy spell. The appraiser may recommend that concrete paint be used which will increase the resistance to moisture seepage. The attic vents may interest him also because they serve two purposes: summer ventilation as a means of lowering attic temperature to improve comfort in the rooms below, and winter ventilation to remove moisture that works through the ceiling and condenses in the attic space. Doors will be checked to see that they are not sprung or warped. Moulding between the floors and the walls will be checked for quality. A loose fit may show that green lumber was used in construction. The appraiser will inspect the kitchen cabinet work to determine if good lumber was used or scraps that accumulated on the job.

You can use the appraisal forms in Figures 5-5 and 5-6 to acquaint yourself with the details an appraiser may include in a

APPRAISAL REPORT

RESIDENTIAL SURVEYS CO.

Property Address______________________________ City______________

Legal Description______________________________

Applicant______________________________ Phone______________

Site is level ☐ sloping ☐ hillside ☐
at grade ☐ above grade ☐ below grade ☐
__________ Feet.

Neighborhood

	Good	Aver.	Poor
Accessability to Shopping	☐	☐	☐
Accessability to Schools	☐	☐	☐
Accessability to Transportation	☐	☐	☐

Zoning______________________________

Typical Price Range______________________________

Typical age______________ years______

Trends Static ☐ Improving ☐ Declining ☐

Comments______________________________

LOCATION SKETCH

Specify And Indicate Direction To Nearest Cross Blvds By Arrow (NCB)
Indicate North Direction By Arrow (N)

Curbs ☐ Street Paved ☐ Street Lights ☐
Sewers ☐ Sidewalks ☐ Other

IMPROVEMENTS: Total Number of Units____ Proposed____ Existing____

(Staple Here) Bldg.____ CONSTRUCTION DETAIL (circle items applying) Existing / Proposed

Attach separate details for each Bldg. except for identical units and garage.

______Sq.Ft. Age______ Stories______ Rooms______ Bedrms______

EXTERIOR: Stucco; Shingle; Board and Batten; Wood Siding______; Trim______; ______; Gutters; Metal Sash

ROOF: Wood Shingle; Wood Shakes; Comp. Shingle; Built-up Rock; Compo Roll; Flat; Mission Tile

WALLS: Plastered ____rms.; Wallboard ____rms.; Sheetrock ____rms.; Decorated ____rms.; Paneling ____rms.; Exp. Beams ____rms.

EXTRAS: Insulation; Garb. Disp.; Dishwash; Refrig.; Range, Oven; Air Cond.; Fans; 220V Plugs; Intercom

FLOORS: Oak______; Pine______; Lin.______; Tile______; Cork______; Conc. Slab______; Asphalt Tile______; Ply. & Carpet______

HEATING, ETC.: Fireplace; Dual Wall Furn.; Dual Floor Furn.; Gravity Furn.; Forced Air; Radiant; Wall Htrs. G.E.; Panel Ray; Gas Outlets; Water Htr.______

PLUMBING: ______Baths; Tub(s) Toilet(s); Pullman Lavatory; Shower: Tub Stall; Tiled Wallboard; Colored Fixtures; Sink: Formica Tile; Floor: Lin. A/Tile

Condition: Poor______ Fair______ Good______ Best______

PROPERTY RATINGS

Ratings	5 Poor	10 Fair	15 Good	20 Best	Reject
Conformity					
Location					
Salability					
Qual. & Condition					
Future Depreciation					

TOTAL RATING

Probable principle cause of future depreciation:

VALUE ESTIMATES (derived from back of page)

Land__________ Sq. Ft.__________ $__________

Depreciated Replacement Cost of all Improvements__________

COST APPROACH VALUE ☐

SALES COMP. VALUE ☐

CAPITALIZED VALUE ☐

Appraisal Report
Figure 5-5

routine appraisal. They are not typical of any particular lender and do not conform to any particular state or federal endorsements. But they can be used by staff and professional fee appraisers alike, and by you as an aid in helping you

Square Foot Appraisal Form for Use With Residential Cost Handbook

Areas and Unit Cost	A	B	Recap	
1. Compute Total Sq. Area			**Sound Value** of Garage	
2. Select House Sq. Cost				
3. **House Adjustments**			Porch____ Sq. Ft. @______	
4. Roofing			Basement______ Sq. Ft. @______	
5. Floor			†Indicate Existing or Proposed	
6. Heat			**Sound Value** of Bldg. I______ S. ft.___ Units @_____	
7. Insulation			**Sound Value** of Bldg. II______ S. ft.___ Units @_____	
8. Total Sq. Adjustment			**Sound Value** of Bldg. III______ S. ft.___ Units @_____	
9. Plumbing			Yard Improvements	
10. Miscellaneous				
11. Multiply Line 1B x Line 8B			**Total of all Improvements**	
Total				
12. Local Multiplier____x Line 11				
13. Total House Repl. Cost				
Depreciation				
14.				
15.				
16. **Sound Value** of Building				

Note Overall Appearance

Quality	Type	Stories
Fair	Conv.	One
Average	Modern	Two
Good	Rustic	Multiple
V. Good	Multiple	

Income Analysis

__________Units @__________per month__________
__________Units @__________per month__________
__________Units @__________per month__________
__________Units @__________per month__________
__________Units @__________per month__________
Total Monthly Schedule__________
Yearly Schedule__________
__________% Vacancy and Loss__________
Adjusted Gross__________
__________% Yearly Expenses__________
Yearly NET Income__________
__________%Interest on Land__________
Net income Attributable to Bldg. only__________

Rate Factor

Interest __________%
Depreciation __________%
Total __________%

Capitalization

__________ ÷ ______ % = __________
Land __________
Total __________

Appraisal Report
Figure 5-6

determine your opinion of value. Construction loans have a different approach and different appraisal requirements, as is shown in Chapter 6.

Loan Settlement Day

Loan settlement is called loan closing by some lenders. The loan closer goes over the funding items with you, including any deposit of funds required for proration of taxes, fire insurance premiums, and the like. Seldom does the borrower receive the full amount of the mortgage loan. Title searches are required, as mentioned earlier, and any necessary surveys and recording fees. Before the loan closing you should receive an orientation booklet and an itemized statement showing the full amount of the loan, all deductions from that amount, and the net amount available to you. If there is no binding commitment, the mortgage in some states does not become a lien on the land until the date on which the loan actually is paid out to the mortgagor. This "binding commitment" on the part of the lender may be a preset date for availability of the proceeds as shown on the application, the day the title policy is issued or another mutually agreed upon date. If there are no binding commitments, the lien is established on the date of disbursement.

In some states lenders use the escrow method. An escrow agent accepts the note and the mortgage from the borrower and the funds from the lender. He pays over to the borrower the proceeds of the loan when it is found to be a good lien. In this way the funds are placed beyond recall by the lender at the time the mortgage is recorded, and the mortgage is documented as a lien from the date of the mortgage recording. Still other states use an opinion of legal counsel stating that the mortgage is a good lien, rather than a title insurance policy which protects the lender from loss due to a faulty lien.

A Note On The Use Of Forms

As a matter of convenience many lenders prefer to use the comprehensive FNMA form in their equity funding procedures. No government sponsorship may be involved. Existing conventional loans may be refinanced through a new conventional loan at a high LVR, and an existing FHA loan may be refinanced through a new conventional loan at a high LVR, but an existing FHA loan may not be refinanced through a new FHA loan to obtain equity funding, or cash.

6

Preparing For A Construction Loan

Every builder is concerned with two types of borrowing: a short term or *interim* loan that carries the project through most or all of the construction, and a *take out* or permanent loan commitment the new owner will use to pay for the property. Before you begin a survey of the loans available you must prepare yourself for every conceivable question a lender might ask in assessing the amount of risk he takes in granting the construction loan. Examine your project in light of your own assessment of the risks. Prepare yourself to meet the lender's requirements, and you will be much more likely to get the loan. Are you going to do all of the construction yourself? How many people will be involved? What part are you going to furnish in terms of labor and materials? Who will be your architect, or are you going to do without one? Who are your subcontractors and what do the plans look like? You should be prepared to answer these and many other questions.

Site Selection for Marketability

Whether or not you intend to resell the project once construction is finished, give some hard thought to the feasibility of the site for both building and marketing. Rolling land is considered unsuitable for low cost housing construction because of the higher cost of preparing the site. But rolling land may be desirable for higher priced homes. A site must have good drainage. Moderately sloping sites are better than either steep land or flat tableland. A 10% slope, for instance, will increase improvement costs; and any substantial grading work causes settlement and erosion problems. Avoid land that is offered at a very low price because of permanently adverse conditions. Distressed property is seldom a bargain. Raw or undeveloped land in a remote area is difficult to value because it takes more time to develop a market survey of sales in the area. And any lender is taking a chance on furnishing a construction loan on a moderately priced home project in a rough wooded area. His risk is increased, so the costs, points and fees on the loan increase.

Consider the marketability of the site as early as possible. Your building site should be removed from potential hazards such as a deteriorated subdivision, railroads, noisy airports or stables. Avoid smoky or damp areas, offensive odors, rundown commercial uses and nearby industrial areas. Cemeteries and nearby fire hazards should also be avoided. If an airport can not be avoided, it may pay to find out if future expansion plans call for jet aircraft. The highway department can be consulted to find out about future development of freeways and access routes in the area. Direct access to a commercial area may be desirable, but adjacent heavily traveled freeways reduce marketability.

Utilities should be within a reasonable distance so that the added cost of extending water, sanitary or storm sewers to the site is not too great. Electricity and gas hookups are very important. Individual projects should provide on-site sewage disposal only when there is no other way of providing for it. In this regard cooperation with the lender is important. He can work with the state board of health concerning sewage disposal. Your lender will support your building project more readily if you have no plans for septic tanks and, especially, no tile fields. Storm sewers should be well separated from the sanitary sewer. It costs less for maintenance with the use of a separate system. The public system should be used for a connection if at all possible or storm drainage could be engineered for regular drain tributaries. For fresh water supply, lenders would rather see the local water main permanently established in the street than back of the sidewalk or dedicated area for the sidewalk. In undeveloped lots, you should anticipate with the lender and the appraiser the location of the water supply. Remember that soil factors—composition, drainage, landscaping and easement problems—must fit the engineering plans for the site so the project is economically feasible in the lender's eyes. If your plans call for septic tanks, more area is usually required in undeveloped lots. Anticipating this kind of problem may tip the balance in your loan application.

Legal Preparation

Recently the Department of Housing and Urban Development has taken an interest in lot sites or land to be developed into lots by subdivision builders. If a builder offers 50 lots or more, he must now provide a report to the potential buyer on the local zoning ordinances restricting the size of the building on the lots, existing liens or loans against the title, utilities and recreational facilities, public services and their projected costs, access to the site and other conditions. And a nationally recognized consumer report service warns builders that "in some cases you cannot build on the land until the last payment is made and you have title . . ." Remember that unless the land is paid for in cash, no construction loan is considered in some states unless a title insurance policy is furnished, or an abstract of title is provided in other states upon applying for the loan.

Zoning and deed restrictions are important to the builder, and the first step before even considering a site in a particular community is a check into the regulations. Past history of the zoning regulations should be noted, because enforcement problems sometimes make ordinances and maps obsolete. If the regulations are not antiquated or if the lender has not experienced problems with deed restrictions in past applications for loans, then chances are good that they are considered an asset to the community.

The first step you should take before building your project is to determine the ownership status of the land. Unless it is a coincidence, it is doubtful that your lender will already know details necessary for potential building sites you are proposing. Therefore, along with determining the quality of the land suitable for building from a construction standpoint, early in the game you must determine the quality of the existing title. Covenants, reservations and restrictions, if any, are found in the ownership. Consult an attorney concerning the quality of the title. It can be said here that *covenants* in a deed or other instrument are agreements promising performance or non-performance of certain acts or stipulating certain uses or non-uses of the property. These agreements go with the land. *Restrictions* are the prohibition of certain uses of the land which could involve the type of structure, minimum square footage of the building, set-back lines for construction and other technical limitations. *Reservations* are rights retained by the grantor in conveying property to others.

In urban areas many homes are bought with no thought of "mineral rights" as an important reservation since the buyer's only concern is to live on the site as a home. No thought is given to drilling for oil or for geothermal mining, for example. But if there were horse stables adjacent to your proposed site your interest in any possible covenants, restrictions or reservations would no doubt be stimulated. Even if you love the country atmosphere of horse stables your construction lender may see a different view. He must think about the future marketability of the land, both for your building plans and in the event he has to exercise foreclosure to recover his investment. Your investigation may disclose that certain steps are being taken by other buyers or subdividers to buy up adjacent property for development of a tract of houses, or that a new zoning law is being considered. Your lender may then want to take another look at your application, subject to inspection of a preliminary title report or a "bobtailed" abstract as a short form of title analysis. Anticipation of a new zoning law may be what is needed to influence your lender to your way of thinking.

Abstracts may include a history of the original title from the government. City and community plats will be noted if such exist. Study easements and restrictions closely, since these go with the land and, unless certain legal steps are taken, will remain in that status. Zoning ordinances limiting the occupancy or uses of the property may show up, specific improvements may not be allowed in the general area or on the land in question.

In the case of subdividers of homesites, copies of covenants, reservations and restrictions must be supplied to the buyer during the final sales process of the property. Keep in mind that any lending institution reviewing your proposal should be informed that zoning for land adjacent to your proposed site has been checked for conformity. Larger cities have a zoning board or department you can consult. Smaller communities have city engineers. And in some cases the city clerk can help you with zoning questions. Unincorporated areas are county-controlled and usually have simpler zoning problems.

You and your lender should check for protective covenants which may be in your favor. Blanket provisions cover the whole area and not simply separate for each deed. Overall restrictions are probably included on the plat map for the area, whereas any special provisions are part of the individual deed. A group of individual covenants that are separate for each deed can be analyzed and dealt with on an individual basis. Be sure to separate blanket provisions from individual covenants. While establishing protection for the loan institutional lenders may try to influence you as a developer to record reasonable and comprehensive protective covenants of your own when you plan your whole project. You can rely on the lender for good financial and legal advice, as he has his investment to protect. Questions can usually be answered by his legal department.

A Construction Checklist

The details of construction methods, special installations, price and quality of materials are outside the direct scope of this book. But be sure you have asked yourself these questions before you settle on a piece of land, especially if you are applying for a sweat equity loan or are a new builder. And be sure the lender you choose is aware that you have asked them and are well versed in the answers to these and other construction questions. Do not assume that the lender will or should answer these questions, though he may indicate where you can obtain more information than you can get on your own.

1. What are the prevailing styles of architecture in your area, and how do you identify them?
2. What are the construction costs per square foot for buildings similar to those in your plans, in different quality ranges and in comparison with other locations?
3. What are the carpenter's terms for the different wood members in the skeleton framework of various buildings similar to your planned construction?
4. What kinds of finish floor materials are available and in what price, wear and cost of maintenance ranges?
5. What are the materials used in the exterior surfacing of projects similar to yours, and what are their relative initial costs and costs of upkeep?
6. What are the minimum and optimum sizes of rooms by width, length and square footage?
7. What are the varieties of heating arrangements available for the proposed project? What are the initial costs, the costs of fuel, and the BTU output of each?
8. How much and what kind of insulation will be used, and at what cost?
9. What are your plans for termite prevention, dry rot and other fungus and insect hazards?
10. What are the desirable roof pitches, materials and slopes for your project? Can you identify various types of roofs?
11. What are the relative advantages or disadvantages of drywall construction compared to plaster?
12. Windows are classified according to position, purpose, materials and operating mechanisms by what identifications?
13. What materials and mechanical arrangements are available for darkening window areas or preventing excess sun infiltration?

You can prepare for your conference with your lender by knowing the building options open to you, the costs involved, the methods and the drawbacks.

The Builder's Contingency Plan

You have selected the building site and have checked into

local zoning and other preliminaries. The business of setting up a purchase of the site is close at hand. If the loan closing on your refinance is not complete, it is in a pending state, so you prepare a contingency plan. Though the project looks possible in every way, and you are sure that buildable land is available for sale, there is a possibility that your refinance loan request now in the hands of the loan committee may not materialize at all. You are aware of contingency plans made in real estate matters so you decide to make a contingency offer of $20,000 to a landowner through his listing broker. Your contingency offer must contain two specific conditions that the seller must agree to. These conditions are preceded by the words "subject to" in the Contract of Sale and Deposit Receipt. (See Figure 6-1.) One condition is based on the refinance loan on your existing property—the actual disbursement must take place. The second condition refers to whether or not a construction loan can be advanced at all on this particular site—many construction lenders consider some lots unbuildable for any number of reasons. The second condition is tied in with the first, and although it is considered a routine stipulation, it is important for your safety. No seller of land for building purposes should object to it. This second condition also provides time for you to conduct a lender's survey to find the best loan.

These contingency conditions are important. (1) If your refinance lender rejects your pending home loan, then you cannot make the cash offer of $20,000. (2) If you cannot build the project for lack of a construction loan, the site is of no value to you. You may not be able to sell or unload the site if there are problems that a construction lender will not invest in. (3) An escrow holder or a third party to the transaction, such as a lawyer, must follow the instructions of the agreement. The second condition guarantees that your full ownership to the lot will not take place until a mortgage of construction money is also ready to be recorded.

The lack of convenient financing for construction seems to be the biggest hazard to builders or any reinvestor. He can readily tie up a substantial amount of his capital when construction financing is not handled on a guaranteed commitment basis. Even if you successfully obtain your

RESIDENTIAL SURVEYS CO.
CONTRACT OF SALE AND DEPOSIT RECEIPT

______________________________, 19____, Time________ AM/PM

RECEIVED from ______________________________

herein called "BUYER," the sum of ______________________________

______________________________ Dollars ($____________) { Check / Note / Irrevocable Transfer }

as a deposit on account of the following described property situated in the

City of ____________, County of ______________________________ to wit:

Lot ____________, Block ____________, Tract ____________, Book ____________, Page ____________

commonly known as ______________________________

Free and clear of all encumbrances of record except as specified.

SUBJECT TO:

(1) Approval, recording, and disbursement of refinance loan by __________ Savings & Loan
(2) Approval, recording, and disbursement of construction loan by __________ Commercial Bank
(3) General and special taxes, including levies for special assessed districts.
(4) Covenants, conditions, restrictions, reservations, easements and rights of way.

Purchase price ______________________________ Dollars

($____________), payable as follows:

____

____

____

____

____

____

____

and it is hereby agreed that:

(1) Escrow shall be opened, or processing begun not later than ten days after acceptance of this offer at a mutually agreed Title Company or at mutually agreed office of an Attorney at law.
(2) Seller agrees to furnish a complete and merchantable abstract of title, or guarantee policy and to pay or reimburse buyer for necessary expenses incurred for abstract-policy.
(3) The contingencies cover a four months period for loan process, or 120 day escrow.
(4) The moneys deposited by the Buyer, as herein provided, are paid as consideration for the execution of this contract of sale. If Buyer, for any reason, fails to complete this contract of sale, such moneys shall be retained and paid one-half to the Broker and the balance to the Seller. The Broker's share of such moneys shall not, in any event, exceed the full commission herein provided for.
(5) Seller will furnish, at his expense, a structural pest control report by a State Licensed operator showing the accessible portion of dwellings and garages to be free of visible evidence of infestation caused by wood destroying insects, fungi, and/or dry rot.
(6) Payoff of existing loans shall be at Seller's expense, and a new loan, obtained by Buyer, shall be at Buyer's expense except where federal regulations provide otherwise.
(7) Buyer represents that he has made an independent investigation of the above property, including plans for proposed freeways, streets, alleys, and easements, and that he is making this purchase in reliance thereon. He further agrees that he is not acting on any representation of Broker or any associated or employee of Broker, except as is specifically set forth in writing herein.
(8) Seller represents and warrants that said property and improvements conform with all applicable ordinances, laws, zoning regulations, and Deed Restrictions and agrees to save agent harmless from any liability or damages due to incorrect information or misrepresentation by the Seller.
(9) Time is of the essence of this contract, but Broker may, without notice, extend for a period of not to exceed one month the time for the performance of any act hereunder, including the date of closing of escrow (except the time for the acceptance of this offer by Seller).
(10) In consideration for the services of the Broker herein, this offer shall remain in effect for a period of ____________ from date hereof.

______________________________ Broker

______________________________ Broker's address and telephone number

ACCEPTANCE BY SELLER

The undersigned accepts the above offer and agrees to sell the property described herein on the terms and conditions herein set forth, and agrees to pay the above named Broker as commission ____________
Should it become necessary for Broker by an action at law to enforce his right to commission, Seller agrees to pay all costs and attorneys' fees incurred therein.

THE UNDERSIGNED ACKNOWLEDGES RECEIPT OF A COPY OF THIS CONTRACT.

____

______________________________ Seller

______________________________ Seller's Address and Telephone Number

ACCEPTANCE BY BUYER

The undersigned Buyer offers and agrees to buy the above described property on the terms and conditions above stated and acknowledges receipt of a copy of this contract.

____

______________________________ Buyer

______________________________ Buyer's Address and Telephone Number

Contract of Sale and Deposit Receipt
Figure 6-1

refinance loan, you are under no obligation to buy the available land unless a satisfactory loan commitment is made in a four-month period as indicated in the sample Contract of Sale and Deposit Receipt in Figure 6-1. Lenders of construction money will not forward the funds to the escrow agent or third party representative until they are assured that all documents necessary have been signed by both principals. That written assurance is authorization to fund the loan according to the loan instructions sent previously for the borrower to sign. The lender takes precautionary measures in the event some situation arises causing the loan to be frozen in escrow indefinitely—he sets aside a trust account for the funds.

Lending Sources for Construction Loans

The three primary sources of conventional construction loans are savings and loan associations, commercial banks and life insurance companies. Loans from the latter source are nearly always obtained through mortgage bankers who "originate" the loan package.

The difficulty with institutional lenders can come in obtaining a combination of two loans—the *interim* or construction loan and the *take out* or permanent loan for the long term repayment. In recent years a system has been introduced among lenders which "packages" the construction loan with the take-out. This sytem has reduced the processing of two separate loans, as the construction lien and the permanent lien reverse positions. In your lender survey for construction loans, note down the active loan institutions in your area. In some locations, especially near large cities, lenders advertise this combination type of loan as a specialty. Full service banks, however, usually do not grant combined construction and take-out loans. In this case the permanent or take-out loan must be applied for separately. Although it is being done by some builders and developers or subdividers, the use of a construction loan without a guaranteed commitment for still another permanent loan for the take-out is considered both clumsy and many times unnecessary. It also may be very expensive.

A construction loan is for a short term. A builder may risk a short maturity on this type of loan which must be paid off quickly—usually in six months to a year. You need a special survey for this type of financing and for the take out. You can be a well-organized builder and still not be able to obtain a guaranteed, or at least a firm, commitment for a permanent mortgage. A small builder or a sweat equity builder should anticipate unforeseen conditions. Note the conditions favorable to the builder in the Contract of Sale and Deposit Receipt form in Figure 6-1. The builder's first concern is loan feasibility.

Depending on the location of the borrower, funds are available through full-service banks, commercial banks, trust companies and community banks. Full service banks insured by the Federal Deposit Insurance Corporation are federal or state chartered corporations owned by stockholders. They are organized for profit. They have the widest range of services of all lenders and their specialty is assessing the potential of businesses in their area. In your case, this is construction. Full service banks nearly always make a construction loan on an interim basis—that is, for the completion of construction only. This is usually for a six month period. A bank-financed construction loan will carry the provision that the work be done by fully active general contractors—unlike the lenders of a sweat equity loan we will discuss later in which the borrower may do some or all of the construction himself. In full service banks the fees or points are non-competitive, therefore higher than lenders who specialize in housing loans. In the case of construction loans through full service banks, it is up to the borrower to find a take out loan during the construction period. Long term mortgages are rare for banks and even then they prefer the FHA insured loan or the GI guaranteed loan program where the risk is low.

Other sources for mortgage funds on an interim basis are mutual savings banks, operated in 18 states, in which the depositor is favored for mortgage construction loans because he is an "owner" who receives dividends rather than interest on his deposit. Mutual banks are non-profit, specialized savings institutions organized and operated for their depositors with no stockholders. They are administered by trustees and are insured by the Federal Deposit Insurance Corporation or by state deposit insurance funds. If you are building in one of these states, it may pay you to join one of these banks to get your loan. A third possibility is a construction loan through membership in a credit union, an organization with more moderate interest and possibly lower separate service fees.

Credit unions take savings from a membership made up of shareholders rather than depositors. Loans at low cost are made to members who govern themselves on a one member, one vote basis. Credit unions are usually made up of people with similar interests, such as employment, professional or religious groups, trade unions, etc. Credit unions are insured by various state insurance programs and the National Credit Union Association, Washington, D.C.

The last source of loans for housing construction, purchase of existing or newly completed housing or home mortgage refinance, is the savings and loan association. They are the housing finance specialists. They are also known in some areas as building and loan associations, cooperative banks, savings associations and even homestead associations.

Savings and loan associations are specialized financial organizations chartered by the states or the federal government. They operate in all states, and are mutually owned by their shareholders. When an account is opened the saver automatically becomes a member of the association. The depositor is thus a shareholder. You are paid dividends on the money that is placed into your account. The federally chartered association is insured by the Federal Savings and Loan Insurance Corporation (FSLIC). The state chartered organizations can be insured by the FSLIC or they may be insured by state chartered savings account insurance organizations.

Loan Correspondents

Not all mortgage lenders are willing to participate in construction lending. They may have any number of reasons, including policy, charter, or money limitations. And there are lenders who prefer long-term financing portfolios and permanent mortgages on completed buildings. Interim lenders on the other hand do not want their money tied up for long periods. A maximum of one year is usually consistent with this type of loan business which survives on constant turnover.

This is because their main income is derived from the points, commissions and fees that several loans produce as opposed to one loan with a long amortization. The interest rate is based on straight loans per year. However, keep in mind that some interim lenders arrange to sell a mortgage loan to a permanent lender-investor after the completion of the project. The sale may be to a permanent lender with whom the interim lender has a steady working arrangement. Or the construction lender may be the *loan correspondent* for a permanent mortgage lender. In your lender's survey you can determine how much more expense you will incur by using a correspondent construction lender than by dealing directly with an institution. A loan correspondent manages the loan. That is, he takes care of taxes and insurance renewals, collects the payments on the loan, remits payments to the permanent lender, making a charge for the collections and other services that may be required. The interim lender, in contrast, needs a *take-out commitment* by the permanent lender, which states that the permanent lender will buy the mortgage after the completion of construction and the expiration of the lien period (usually four months).

Aside from the loan correspondent take-out/permanent mortgagee combination, there is another system in which the interim lender assigns the mortgage to a permanent lender that the builder himself arranges. Under this system a builder obtains from some permanent lender the commitment to make a new mortgage loan upon project completion. The proceeds of the new mortgage are then used to pay off the construction mortgage. In this type of take-out commitment, the builder-mortgagor applies to a permanent lender as the first step in the project financing process. The permanent lender then makes the commitment to the mortgagor, *not* the interim construction lender. The builder-mortgagor then assigns this commitment to the construction lender. Under this system, only then would the construction lender make a firm commitment.

Some savings and loan associations make available special construction loan services to develop and promote new business and construction. They offer sample plans and specifications with building material exhibits for builders. Frequently a construction library and architectural advisors are available; in other instances an accounting service is provided for builders in the area. You can use these services whether you are a large builder or an owner intending to build. These services are merely promotional, and when available are part of the overall cost of obtaining the loan. It is not an extra item of expense separated specifically from a construction loan but may be absorbed into the origination fee.

Basic Construction Loan Process

A construction loan process must be adapted to local contracts, lien laws and other legal conditions subject to various state laws. Any lending institution has its attorney study the statutes affecting construction loans so that the loan is adequately protected.This list may need adjustments to make it conform to provisions in a particular state. But these are the basic steps you will take in obtaining any construction loan.

1. Lender receives the loan application which includes
   - A. Local information about the builder-borrower
   - B. Location of the proposed project
   - C. Plans, specifications, plot plan and cost estimate
   - D. Name of corporate executive or builder
2. Lender makes a credit investigation and determines the borrower's reliability.
3. Lender appraises the site of the proposed project based on the plans and specs.
4. Loan is submitted to loan committee for approval.
5. Loan is closed, including:
   - A. Building contract
   - B. Note and mortgage
   - C. Loan contract
   - D. Preconstruction affidavit
   - E. Receiving borrower's equity
   - F. Ordering fire and hazard insurance
6. Mortgage papers are recorded, builder's contract is recorded if customary.
7. Site is inspected to determine whether or not materials are on the site.
8. Lender notifies the owner and contractor to begin construction, accounting record system is set up.

9. Lender schedules advances to the borrower or builder as work proceeds.

10. Lender schedules periodic inspections, writes final report of his inspection of the completed construction and makes final disbursement of funds.

Sweat Equity Loans

When the site owner's broker reveals the acceptance of the contingency deal, both the buyer and the seller of the site need a construction loan as soon as possible. If the builder is inexperienced or planning his project as housing for himself, he may not be able to obtain a construction loan as easily as big builders. He needs a sweat equity lender. *Sweat equity* is an allowable credit from a lender to a borrower in lieu of part of the construction cost for labor. Sweat equity lenders, most of whom are savings and loan organizations, may decrease the required cash deposit if the owner-builder furnishes some of his own labor in the project. The interim funds are advanced for construction of building projects on a small scale with as little risk of project failure as possible. They differ from professional builder's construction loans because a professional is usually less risk due to his reputation. Many sweat equity loans are made in communities where the officers of the lending group are familiar with the townspeople. Usually the fees or points are comparatively high, 4-6% or more. The usual rule for savings associations concerning the maximum cost the applicant can afford to pay for the house, including sweat equity, is three times his net income. The monthly payment on the package, including interim and take-out, is around ¼ of the mortgagor's net income. The amortization can be tailored to fit this rule of thumb.

Although the lender will expect you to contribute some part of the construction cost in a down payment, rather than merely owning the land or building site outright, this is not an absolute requirement. But it decreases the lender's risk if you have cash in your own project. An extra point or two or a slightly higher rate of interest may be charged you without it, depending upon the lender's policy. Even a down payment of 5% can show your willingness to cooperate. A mortgagor's completion bond is seldom required; but if it is, you must furnish it. Construction lenders often prefer instead to depend upon their own personnel to see that all work goes along with the schedule and rely on a *completion bond waiver* as shown in Figure 6-2. This is especially preferred if a relative will underwrite the construction loan. This is added security for the lender if the owner-builder is relatively inexperienced and cannot contribute a substantial amount of money out of pocket toward the construction.

Preparing for a Sweat Equity Loan

When you approach the lender, be prepared to give him an indication of your degree of competence in construction and your financial situation. The applicant for this type of loan is rarely a building professional; most often he is building his own residence with the help of relatives and friends or for resale income for his own small-scale business. The ideal sweat equity borrower is a carpenter, plumber, mason, or electrician. Better yet, he possesses all of these skills. The lender may not expect you to be all these things; your assurance to him that some of the work is to be subcontracted out should count in your favor. Even rudimentary skills contribute to the lender's estimation of your ability, if you can show that you can improve and expand your skills as the project progresses.

The sweat equity lender will ask you if you intend to occupy the house after construction or whether the project is purely for profit. Your answer to this question could be the difference between getting the loan and having it turned down. Many lenders will not assume the risk of sweat equity, and will flatly refuse your application for a combination interim/takeout loan package. The pitfalls for him are always there—at least in his imagined picture of you losing interest, finding yourself over your head in the project, lacking experience in some vital construction area halfway through the building. But if you can convince him that you have some mechanical aptitude or construction background and have the stamina to finish what you start, he may bend in your favor. He looks for your ability to stick to the job until completion, a sense of responsibility and a proven record of quality work. He must be convinced you can build the house yourself with the help of friends,

LIEN AND COMPLETION BOND WAIVER

ALL CITY BANK Loan No. ____________

____________________ State

____________________ Office ____________________, 19___

Gentlemen:

This is with reference to your loan to me secured by a deed of trust on the property hereinafter described. I hand you an executed copy of building contract dated between myself as owner and ______________________________ as contractor, and a copy of the **plans and specifications** referred to therein, identified by my signature, for the construction of improvements to cost $__________ on property owned by me and legally described as:

I have made a careful investigation of the financial and moral responsibility of said contractor and I am satisfied that this responsibility is such as to justify the waiving of the contractor's completion and lien bond. Consequently, I hereby request that you consent to the construction proceedings without requiring the contractor to furnish said bond.

In consideration of your making the loan above referred to, and of your complying with my request that the contractor's bond be waived, I hereby undertake and agree:

1) That construction of said improvement shall be commenced within thirty days from the date of this instrument, shall thereafter be diligently prosecuted, and shall be fully completed within________days from the date construction has commenced according to the requirements, rules and regulations of the Federal Housing Administration, the Veterans Administration, and/or All City Bank.

2) That said improvements shall be constructed and completed in strict accordance with the plans and specifications referred to in the above mentioned contract now on file with you.

3) Should the contractor fail to commence or complete the building within the time and in the manner herein provided, I hereby agree that I will commence and complete the same at my own cost and expense.

4) That I will protect and indemnify you against any loss sustained by you by reason of your having waived a lien and completion bond herein referred to.

5) In the event that any liens should be filed against this property in connection with the erection of this building or any attachments or executions be filed against building funds being held by you, I will pay on your demand the amount necessary for the release of such liens, attachments or executions in order that the building may be completed.

6) That the obligations and agreements herein contained shall in no event be construed as altering, amending, diminishing, or affecting in any manner whatsoever the covenants, obligations, and agreements contained in the deed of trust executed by us, securing the loan herein referred to.

7) That we will purchase Fire Insurance coverage for $____________and deliver the same to the Bank

____
Owner

____
Owner

Completion bond waiver
Figure 6-2

neighbors, relatives or skilled and semiskilled labor.

Can your wife assist in the planning and building of the project? This is important to him whether or not the two of you are operating as a team.

The lender expects you to furnish clear plans and specifications for the accurate estimation of costs. Plan companies can supply standard plans to use. Do not rely on thumbnail sketches or rough drawings with general outlines and doubtful dimensions. At the same time, your plans and specs should reflect a moderate project rather than multiple bedrooms and more bathrooms than necessary for your needs. Additions can be made in later years. If you are to subcontract some of the work yourself you should include several bids for certain kinds of work such as plastering or concrete. You may lose much time and money if the local building codes require licensed plumbers for the plumbing work, for instance, and you have begun much of this specialized work yourself. When you bid out some work, your lender will expect you to obtain bids for that work under a guarantee—the price must be reliable or it is not a dependable subcontractor's bid. The wise builder shops around for two or three bids; but in smaller towns the lender usually knows the reliable contractors, and he would rather have a cooperative spirit in a mutual selection than see you taken advantage of in a cut-rate deal you selected alone. If you are inexperienced in collecting bids, let the lender guide you. He will expect these bids to be reasonable in the light of market conditions. If a bid is too low the contractor may not be able to complete the project. He may ask for more funds, and if it appears that they are not available, he may walk out without completing his phase of the work. If the bid is too high, the borrower would have an unhappy investment and the lender would be in an insecure position. All bids should be accumulated in your estimated cost of the entire project.

If the lender seriously considers making the combination interim and take-out loan he will begin an investigation of the building as a lending risk. This includes a credit rating, current debts, weekly or monthly income stubs or statement of business income. This may include an analysis of character and reliability gathered from friends and business associates.

U. S. DEPARTMENT OF HOUSING AND URBAN DEVELOPMENT
FEDERAL HOUSING ADMINISTRATION

2. FHA Case No. ▲

1. SPECIAL PROCESSING ▲
1. ☐ Veteran 2. ☐ Assistance Payment 4. ☐

DATA PAGE
MORTGAGE TO BE INSURED UNDER
☐ SEC. 203(b) ☐ SEC.

3

3. PROPERTY ADDRESS

4. MORTGAGORS/BORROWERS:
Mtgor. ______ Sex ______ Age ______
Co-Mtgor. ______ Sex ______ Age ______
Address ______

5.

Married ▲ Yrs. No. of Dependents ▲ Ages ▲
Co-Mortgagor(s) Sex ▲ Age(s) ▲
(Check one) ☐ White, not of Hispanic origin ☐ Asian or Pacific Islander
☐ Black, not of Hispanic origin ☐ Hispanic
☐ American Indian or Alaskan Native ☐ Other

6. MORTGAGE APPLIED FOR →	Mortgage Amount	Interest Rate	No. of Months	Monthly Payment Principal & Interest
	$	%		$

7.
PURPOSE OF LOAN: ▲ (1) ☐ Finance Constr. on Own Land (2) ☐ Finance Purchase (3) ☐ Refinance Exist. Loan (4) ☐ Finance Impr. to Exist. Prop. (5) ☐ Other
MORTGAGOR WILL BE: ▲ (1) ☐ Occupant (2) ☐ Landlord (3) ☐ Builder (4) ☐ Escrow Commit. Mortgagor

8. **ASSETS**
Cash accounts ______ $ ______
Marketable securities ______
Other (explain) ______
OTHER ASSETS (A) TOTAL ▲$ ______
Cash deposit on purchase ______
Other (explain) ______
(B) TOTAL $ ______

9. **LIABILITIES** Monthly Payt. Unpd. Bal.
Automobile $ ______ $ ______
Debts, other Real Estate ______
Life Insurance Loans ______
Notes payable ______
Credit Union ______
Retail accounts ______
NAME ACCOUNT NO.
(If additional space needed, attach schedule) TOTAL $ ______ ▲$ ______

10. **EMPLOYMENT**
Mortgagor's occupation ______
Employer's name & address ______
______ years employed ______
Co-Mtgor. occupation ______
Employer's name & address ______
______ years employed ______

11. **MONTHLY INCOME**
EFFECTIVE INCOME MONTHLY INCOME
▲$ ______ . . Mortgagor's base pay . . ▲$ ______
______ . . . . Other Earnings . . . . ______
▲ ______ . Co-Mortgagor's base pay . ▲ ______
______ . . . . Other Earnings . . . . ______
______ . Income, other Real Estate . ______
______ . . . . Other. . . . . . . ______
▲ ______ . . . . . . TOTAL . . . . . . ▲$ ______
______ . Less Federal Income Tax .
▲$ NET EFFECTIVE INCOME

12. **SETTLEMENT REQUIREMENTS**
(a) Existing debt (Refinancing only) . . . .$ ______ $ ______
(b) Sale price (Realty only) . . . . ______ ▲ ______
(c) Repairs & Improvements . . . . ______
(d) Closing Costs . . . . ______ ▲ ______
(e) TOTAL (a + b + c + d) Acquisition cost ______ ▲ ______
(f) Mortgage amount . . . . ______
(g) Mortgagor's required investment(e-f). . ______
(h) Prepayable expenses . . . . ______
(i) Non-realty & other items . . . . ______
(j) TOTAL REQUIREMENTS (g + h + i) ______
(k) Amt. pd. ☐ cash ☐ Other (explain) . . ______
(l) Amt. to be pd. ☐ cash ☐ Other (explain) ______
(m) Tot. assets available for closing (8)(A) $ ______ $ ______

13. **FUTURE MONTHLY PAYMENTS**
(a) Principal & Interest . . . .$ ______ $ ______
(b) FHA Mortgage Insurance Premium . ______
(c) Ground rent (Leasehold only) . . . . . ______
(d) TOTAL DEBT SERVICE (a + b + c) ______
(e) Hazard Insurance. . . . . ______
(f) Taxes, special assessments . . . . . ______ ▲ ______
(g) TOTAL MTG. PAYT. (d + e + f). . ______ ▲ ______
(h) Maintenance & Common Expense . . ______ ▲ ______
(i) Heat & utilities . . . . ______
(j) TOTAL HSG. EXPENSE (g + h + i) ______ ▲ ______
(k) Other recurring charges (explain). . . ______
(l) TOTAL FIXED PAYT. (j + k) . . . .$ ______ ▲$ ______

14. **PREVIOUS MONTHLY HOUSING EXPENSE**
Mortgage payment or rent . . . . $ ______
Hazard Insurance. ______
Taxes, special assessments . . . . ______
Maintenance . . . . ______
Heat & Utilities . . . . ______
Other (explain) . . . . ______
TOTAL ▲$ ______

15. **PREVIOUS MONTHLY FIXED CHARGES**
Federal, State & Local income taxes . . . . $ ______
Prem. for $ ______ Life Insurance. . . . . ______
Social Security & Retirement Payments . . . . ______
Installment account payments . . . . ______
Operating Expenses, other Real Estate. . . . . ______
Other (explain) . . . . ______
TOTAL $ ______

Figure 6-3

A sweat equity loan granted to an owner-builder or a loan made to an individual who is contracting to build his own housing may be less risky to the lender than a loan made to a builder without a prospective buyer. Although either type of borrower of building funds will be thoroughly analyzed for his

16. Do you own other Real Estate ☐ Yes ☐ No Is it to be sold ☐ Yes ☐ No FHA mortgage ☐ Yes ☐ No Sales Price $ Orig-Mtg. Amt. $
Unpaid Bal. $ Address Lender

17. **RATIOS:** Loan to Value %: Term to Remain. Econ. Life %: Total Payt. to Rental Value %: Debt Serv. to Rent Inc. %

18. **MORTGAGOR RATING**__________
Credit Characteristics__________ Motivating Interest in Ownership__________ Importance of Monetary Interest__________
Adequacy of Available Assets__________ Stability of Effective Income__________ Adequacy of Effective Income__________

Remarks:

Examiner: Reviewer: Date: 19

Figure 6-4

financial condition, the marketability of the finished product may appear more important where the professional builder may depend upon resale. This is why most lenders, unless acquainted with the professional builder, may require a separate financial statement as well as a list of previous customers. It stands to reason that the individual owner-builder, although not in the business of construction for profit, may not have to undergo the same formalities. This financial statement is required of a professional builder for profit and covers his cash assets, any currently owned real estate including improved or unimproved sites, equipment owned, and stocks, bonds or any marketable securities. The form shown in Figures 6-3 and 6-4 also indicates all liabilities of the developer-builder as well as tax and current mortgage payment information. Primarily, it is a *contractor's financial statement,* called FHA-2900-3 (4-78).

Survey records are available to lenders, title companies or persons anticipating a vested interest in construction sites. But the legal descriptions are reaffirmed by special survey work. Before the loan is considered complete for construction, these special surveys for plot site must be made to avoid possible encroachment, violation of building codes and the like. Shortly after the forms are installed a recheck of foundation wall to the property line will verify compliance with the local building codes.

Plans and specifications are now analyzed very conservatively by the lender. With the description of the work to be done, the material to be furnished, the time payment system and its proposed schedule, the labor costs and the architectural procedures studied and compared, a uniform estimate can be established. If for any reason the lender finds large deviations in a cost projection based on subcontractor's bid, then further

ALL CITY BANK

CONTRACT DECLARATION (subcontractors)

Date:____________________

To:____.
Our files indicate that your organization has contracted with ________________________for the purpose of supplying labor and material in the construction of a building owned by ____.
The above named general contractor indicates that you are to furnish ____.
We ask your cooperation in returning this letter with the following information which will facilitate final disbursement details.

Amount of your contract:____________________________.
Amount received to date:____________________________.
Amount presently due you:___________________________.
Date:____.
Signature:______________________________________.

Our pay-out department draws all checks directly to the supplier or subcontractor upon receipt of general contractor's order for disbursement. Payments are deducted from the individual amounts set up with new balances brought forward, as per pay-out schedule.

Yours very sincerely,

____

All City Bank

Figure 6-5

investigation of your application will be warranted. A complete breakdown of material costs, the exact contract price and labor to be supplied, as well as a description of the particular order issued by the general contractor, is now obtained from the subs. This may appear as a form letter from the lender requesting an accurate statement. See Figure 6-5.

Separate Contracts with Common Cause

There are three separate but basic contracts. Each is dependent upon the other. An ideal loan package is incomplete without these three documents.

Building construction contract Although this contract is a mutual agreement between the property owner and the contractor, it is completely separate from any contract you may have with a lender. Three copies of the plans and specifications must be signed by the owner-builder and his subcontractors. A standard form for such a building construction contract is shown in Figures 6-6, 6-7, 6-8 and 6-9. Although different forms are found in different states or areas, the important points that should be included in this contract are (1) the dollar amount covering the work to be done, (2) the terms and conditions for the payment of the job, (3) the contractor's liabilities to the job and to his workmen, (4) a time limit on completion of the work, and (5) a clause covering any penalty in case the completion date is exceeded. The owner and the contractor each keep a copy of the contract and a signed copy of the attached plans and specifications. A third copy of the contract is made available to the mortgage lender, with signatures on the attached plans and specifications.

Construction loan agreement This is a separate agreement between you and the lender, and relies on the valid building construction contract. The construction loan agreement is signed by you as the owner-builder, the contractor, and an officer of the lending firm. It is signed at the same time as the mortgage. When you sign the construction loan agreement you also sign a clause giving the lender the right to supervise the building project. Keep this in mind when you work with him during construction. If he thinks it is necessary, he can step in to relet every contract on the job with your approval. He reserves this flexibility to protect his funds and to assure a completion date. But the lender does not

BUILDING CONSTRUCTION CONTRACT

THIS AGREEMENT, made this fifteenth day of March, 19 77,

between Jack B. Nimble, hereinafter

called Owner, whose address is 66 Magnolia Way, Walnut Creek,

and Frank Nayler and Associates, hereinafter

called Contractor, License No. 72491 AZ, whose address is 12 Broadway, Walnut Creek

In consideration of the covenants and agreements herein contained, the parties hereto agree as follows:

1. Contractor agrees to construct and complete in a good, workmanlike and substantial manner, upon the real property hereinafter described, furnishing all labor, materials, tools and equipment therefor, a

Single family residence and attatched garage

(hereinafter called the structure, whether one or more buildings or improvements), upon the following described real property:

Lot 3 Block 6 Walnut Creek Tract

Book 66 Page 389

2. The structure is to be constructed and completed in strict conformance with plans and specifications for the same signed by the parties hereto, a copy of which plans and specifications have been filed with

1. Owner 2. Contractor 3. Mortgagee

Hereinafter referred to as Lien Holder. If no Lien Holder is named herein all reference to same in this contract is to be disregarded.

The structure is also to be constructed and completed in strict compliance with all laws, ordinances, rules and regulations of competent public authority, and Contractor is to apply for and obtain all required permits, paying all fees therefor, and all other fees required by such public authority.

3. In consideration of the covenants and agreements hereof being strictly performed and kept by Contractor, including the supplying of all labor, materials and services required by this Contract, and the construction and completion of the structure, Owner agrees to pay to Contractor the sum of $ 30,000, in installments as follows: In accordance with the disbursement program as formed by owner and mortgagee or first lien holder.

Figure 6-6

commit himself for the quality of the workmanship. Yours and the lender's rights are fully set out in the construction loan agreement. See Figure 6-10, 6-11, 6-12, and 6-12A.

4. The Contractor agrees to commence work hereunder within 10 days after receipt of written notice from the owner and Lien Holder so to do, to prosecute said work thereafter diligently and continuously to completion, and in any and all events to complete the same within 120 days after commencement of work as aforesaid, subject to such delays as are permissible under paragraph 10 hereinbelow. In no event shall the Contractor commence said work or place any materials on the site thereof prior to receipt of such notice from the owner.

5. Contractor shall pay promptly all valid bills and charges for material, labor or otherwise in connection with or arising out of the construction of said structure and will hold Owner of the property free and harmless against all liens and claims of lien for labor and material, or either of them, filed against the property or any part thereof, and from and against all expense and liability in connection therewith, including, but not limited to, court costs and attorney's fees resulting or arising therefrom. Should any liens or claims of lien be filed for record against the property, or should Owner receive notice of any unpaid bill or charge in connection with the construction, Contractor shall forthwith either pay and discharge the same and cause the same to be released of record, or shall furnish Owner with proper indemnity either by satisfactory corporate surety bond or satisfactory title policy, which indemnity shall also be subject to approval of Lien Holder.

6. Contractor shall, if requested, before being entitled to receive the second or any subsequent payment herewith, furnish to Owner all bills paid to that date, properly receipted and identified, covering work done upon and materials furnished for said structure and showing an expenditure of an amount not less than the total of all previous payments made hereunder by Owner to Contractor.

7. The plans and specifications are intended to supplement each other, so that any works exhibited in either and not mentioned in the other are to be executed the same as if they were mentioned and set forth in both.

8. Should the Owner at any time during the progress of the work request any modification, alterations or deviations in, additions to, or omissions from, this contract or the plans or specifications, he shall be at liberty to do so, and the same shall in no way affect or make void this contract; but the amount thereof shall be added to or deducted from the amount of the contract price aforesaid, as the case may be, by a fair and reasonable valuation, based upon the actual cost of labor and materials plus 10% profit to the Contractor. And this contract shall be held to be completed when the work is finished in accordance with the original plans as amended or modified by such changes, whatever may be the nature or extent thereof. The rule of practice to be observed in fulfillment of this paragraph shall be that upon the demand of either the Owner or the Contractor, the character and valuation of any or all changes, omissions or extra work shall be agreed upon and fixed in writing, signed by the Owner and the Contractor, prior to execution. Where the alterations, deviations, additions, or omissions from the said plans or specifications require the written approval of the Lien Holder the Owner will secure said written approval. Provided however that the Contractor is not deemed to have waived his right to compensation for extra work if the same is not provided for in writing.

9. Should Contractor, at any time during the progress of the work, refuse or neglect to supply sufficient material or workmen for the expeditious progress of said work, Owner may, upon giving three days' notice in writing to Contractor, by registered mail, (a copy of which shall be furnished to aforesaid Lien Holder), provide the necessary material and workmen to finish the said work and may enter upon the premises for such purpose and complete said work, the expense thereof shall be deducted from the said contract price, or if the total cost of the work to Owner exceeds the contract price, Contractor shall pay to Owner upon demand the amount of such excess in addition to any and all other damages to which Owner may be entitled. In such event Owner may take possession of all materials and appliances belonging to Contractor upon or adjacent to the premises upon which said work is being performed and may use the same in the completion of said work.

10. The time during which the Contractor is delayed in said work by (a) the acts of Owner or his agents or employees or those claiming under agreement with or grant from Owner, or by (b) the Acts of God which Contractor could not have reasonably foreseen and provided against, or by (c) stormy or inclement weather which necessarily delays the work or by (d) any strikes, boycotts or like obstructive actions by employees or labor organizations and which are beyond the control of Contractor and which he cannot reasonably overcome, shall be added to the time for completion by a fair and reasonable allowance.

11. The Contractor shall not be responsible for any damage occasioned by the Owner or Owner's agent, Acts of God, earthquake, or other causes beyond the control of Contractor, unless otherwise herein provided or unless he is obligated by the terms hereof to provide insurance against such hazard or hazards. It is understood and agreed that Contractor, before incurring any other expense or purchasing any other materials for this work, shall proceed with the foundation work and that if, at the time of excavation therefor, the Contractor finds that extra foundation

Figure 6-7

work is required he shall so notify the Owner, and Owner shall at that time have the right and option to immediately cancel and terminate the within contract or to deposit the estimated cost of the required extra foundation work with the Lending Agency, if there be one named herein, or add such amount to the contract funds wherever same are at that time deposited; it being agreed that in the event of a cancellation the Contractor shall be paid his actual costs of the work done to the time of cancellation. In computing said costs building permit fees, insurance and such financing and title charges as are not refundable shall be included; but supervision time, office overhead and profit are not to be included.

12. No payment hereunder nor occupancy of said improvements or any part thereof shall be construed as an acceptance of any work done up to the time of such payment or occupancy, except such items as are plainly evident to anyone not experienced in construction work, but the entire work is to be subject to the inspection and approval of Owner at the time when it shall be claimed by Contractor that the work has been completed. At the completion of the work should there be any minor items in question, or to be adjusted, i. e., items not of a substantial nature, Owner may withhold from the payment then due a sum equal to twice the fairly estimated amount of money required to cover said items or involved by such adjustments, and shall pay the difference to the Contractor. It is understood and agreed that the acceptance of any works by the Owner shall not be construed to be an acceptance by aforesaid Lien Holder who is not a party hereto.

13. Owner agrees to sign and file for record within ten days after the completion and acceptance of said work a notice of completion (a copy thereof to be deposited with aforesaid Lien Holder at least forty-eight hours prior to such recording), and Contractor agrees upon receipt of final payment, to release the said work and property from any and all claims that may have accrued against same by reason of said construction. If the Contractor faithfully performs the obligations of this Contract on his part to be performed, he shall have the right to refuse to permit occupancy of the structure by the Owner or Owner's agent until Contractor has received the payment, if any, due hereunder at completion of construction, less such amounts as may be retained pursuant to mutual agreement of the Owner and Contractor under the provisions of the preceding paragraph hereof. Said Lien Holder has the right to make its own decisions as to the completion of any work, independent of the parties hereto.

14. Owner agrees to procure at its own expense and prior to the commencement of any work hereunder, fire insurance with course of construction clause and waiver of fallen building clause attached in a sum equal to the total cost of said improvements as set forth in paragraph 3 hereof, with loss, if any, payable to any mortgagee or beneficiary, such insurance to be written to protect the Owner and the Contractor, as their interests may appear, and should Owner fail so to do, Contractor may procure such insurance, but is not required to do so, and Owner agrees on demand to reimburse Contractor in cash for the cost thereof.

15. Contractor shall at his own expense carry all workmen's compensation insurance and public liability insurance necessary for the full protection of Contractor and Owner during the progress of the work. Certificates of such insurance shall be filed with Owner and with said Lien Holder if Owner so requires, and shall be subject to the approval of both of them as to adequacy of protection.

16. Any controversy or claim arising out of or relating to this contract, or the breach thereof shall be settled by arbitration in accordance with the Rules of the American Arbitration Association, and judgment upon the award rendered by the Arbitrator(s) may be entered in any Court having jurisdiction thereof. Any such award may include costs and reasonable attorney's fees as may be directed by the Arbitrator(s).

17. Should either party hereto bring suit in court to enforce the terms hereof any judgment awarded shall include court costs and reasonable attorney's fees to the successful party.

18. Upon the completion of the work the Contractor agrees to remove all debris and surplus materials from Owner's said property (including underarea of structure) and leave said property in a neat and broom clean condition.

19. The Contractor shall not assign or transfer this contract without first obtaining Owner's consent in writing.

20. The aforesaid Lien Holder is not a party to this Contract and is not bound or obligated by or under any of the terms hereof, except item 13.

21. Time is of the essence of this Contract as to both parties hereto.

Figure 6-8

Contractors are required by law to be licensed and regulated by the Contractors' State License Board. Any question concerning a contractor may be referred to the registrar of the board whose address is:

Contractors' State License Board

"NOTICE TO OWNER"

Under the Mechanics' Lien Law, any contractor, subcontractor, laborer, materialman or other person who helps to improve your property and is not paid for his labor, services or material, has a right to enforce his claim against your property. This means that, after a court hearing, your property could be sold by a court officer and the proceeds of the sale used to satisfy the indebtedness. This can happen even if you have paid your own contractor in full, if the subcontractor, laborer, or supplier remains unpaid.

Under the law you may protect yourself against such claims by filing, before commencing such work of improvement, an original contract for the work of improvement or a modification thereof, in the office of the county recorder of the county where the property is situated and requiring that a contractor's payment bond be recorded in such office. Said bond shall be in an amount not less than fifty percent (50%) of the contract price and shall, in addition to any conditions for the performance of the contract, be conditioned for the payment in full of the claims of all persons furnishing labor, services, equipment or materials for the work described in said contract. The above wording is not typical of any specific state law.

IN WITNESS WHEREOF, the said parties hereunto set their hands the day and year first above written.

Jack B Nimble

A Frank Nayler

Figure 6-9

ALL CITY BANK

BUILDING LOAN AGREEMENT AND ASSIGNMENT OF ACCOUNT

This agreement is executed by the undersigned owner, or owners, hereinafter called "Borrower", for the purpose of obtaining a construction loan from All City Bank, a Corporation, hereinafter called "Bank", which loan is to be evidenced by a Note of Borrower for $__________ dated __________ in favor of Bank and is to be secured by a Deed of Trust on real property (including all buildings and improvements now or hereafter constructed thereon) owned by Borrower situated in the County of__________, State of__________ and described as follows:

The net proceeds of this loan, upon recordation of the Deed of Trust and receipt by Bank of evidence that Title Insurer shall have issued or agreed to issue a title policy required by Bank and naming Bank as insured to the extent of the loan amount, are to be deposited in a special non-interest bearing account with Bank entitled Building Loan Account, and the sum of $ __________ (provided by Borrower) is to be deposited in the same account. Borrower agrees that the deposit of said loan proceeds into said account shall be conclusively deemed to be full and complete consideration for said Note and Deed of Trust and that such consideration shall be deemed to have been fully paid to Borrower. Subject to the provisions of this Agreement, Borrower hereby irrevocably assigns to Bank as security for the obligations secured by said Deed of Trust and the due performance of the terms of this Agreement by Borrower and for any other joint and/or several obligations of Borrower to Bank, all of the rights, title and interest of Borrower in and to said account and to all funds placed and to be placed therein. Borrower waives any right to or claim against funds in said account other than to have the same disbursed by Bank in accordance with this Agreement, which disbursement the bank, upon its acceptance hereof, agrees to make for the purposes and upon the conditions set forth herein. It is mutually understood that all disbursements from said funds shall be deemed to have been made first from the funds provided by Borrower and until said sums are exhausted, the funds remaining in the account shall be conclusively deemed to be loan funds advanced by Bank. Borrower further agrees as follows:

1. If required by Bank as indicated by an "X" in the box opposite the required item, Bank shall have received prior to recording of the Deed of Trust and prior to disbursement of any proceeds of this loan:

☐ a) Executed Continuing Guaranty of __________

☐ b) A performance bond naming Bank as co-obligee, and labor and material payment bond, in a penal sum equal to the amount of the General Contract, or if none, then in such amounts as Bank may require, in form and content meeting the statutory requirements

☐ c) Executed Security Agreement

☐ d. Original executed Major Leases and an Assignment thereof to Bank on Bank's form.

☐ e) A copy of the applicable zoning ordinances certified by an appropriate official to be true, complete and up to date.

☐ f) Letters from local utility companies or local authoritiy stating that electric, gas, sewer, water and telephone facilities will be available to the property upon completion of the improvements.

☐ g) A Guaranty of Completion satisfactory to Bank duly executed by __________

☐ h) Subordination Agreements executed in recordable form

☐ i) Executed Assignment of Rents.

☐ j) Assignment to Bank of all leasehold interest held by Borrower on the above described property.

☐ k) Other collateral consisting of __________

Figure 6-10

2. To furnish Bank, prior to recording of the Deed of Trust, and prior to the disbursement of any proceeds of this loan: a) the executed Note, b) the executed Trust Deed, c) the Building Permit and any other authorization, if any, which may be required from the Local Authority or Government Authority, d) the Financial Statements of Borrower, e) the first page of a copy of the Plans and Specifications signed and all other pages initialed by Borrower, General Contractor, Permanent Lender and/or Participating Lender, if any, and any tenant under a Major Lease if required by Bank, f) a copy of the General Contract, if any, g) a copy of the Cost Breakdown or detailed Cost Estimate signed by Borrower and by General Contractor, if any, h) if Borrower is a corporation, then a certified Resolution of Board of Directors of Borrower authorizing the consummation of the transactions contemplated hereby and providing for the execution of a written direction of payment if Loan proceeds are to be paid to a person other than Borrower, i) if Borrower is a General Partnership or Joint Venture, the General Partnership or Joint Venture Agreement, j) if Borrower is a Limited Partnership, the Limited Partnership Agreement and the Certificate of Limited Partnership, if requested by Bank k) if Borrower is a Trust, a copy of the Trust Indenture, l) if Borrower is an Estate, a copy of the Letters Testementary and a Court Order authorizing the borrowing.

3. No work of any character will be commenced nor will any materials be delivered upon or near said real property until the Borrower shall have received notice from Bank authorizing the commencement of work, or until a proper and satisfactory indemnity agreement has been arranged with the Title Insurer. Should any work be commenced or any materials be delivered upon or near said real property before receipt by Borrower of such notice to commence, without said indemnity agreement being arranged, Bank may apply to the indebtedness secured by said Deed of Trust as much of the funds in said loan account as may be required to satisfy said indebtedness and to pay all expenses incurred in connection with this loan transaction.

4. Funds held in the Building Loan Account shall be disbursed in accordance with the Inspection and Disbursement Schedule (Exhibit "A" hereto) and subject to the following conditions:

a) Funds held but not necessarily for construction shall be disbursed at the Banks option and for the purpose so held.

b) Regular and last disbursements of loan funds shall require advice from the Bank's Inspection Department to the effect that improvements have been completed substantially in accordance with plans, specifications and cost estimates.

c) Upon completion of foundations, Borrower shall have notified Bank of said completion and Title Insurer shall have issued or agreed to issue its foundation indorsement assuring Bank that the foundation is constructed wholly within the boundaries of the property and does not encroach on any easements or set-back limits.

d) The last disbursement for construction purposes shall require receipt by Bank of: 1) Evidence that Borrower has filed the Notice of Completion of the improvements necessary to establish commencement of the shortest statutory period for the filing of mechanic's and materialmen's liens. 2) Evidence that Title Insurer has issued or has agreed to issue its ALTA rewrite policy, or its 101 indorsement and/or such other indorsements satisfactory to Bank. 3) Evidence that no unpaid claims or other liens are on file with Bank against any of said funds. 4) If requested by Bank, in lieu of 1) 2) and 3) heretofore, indemnification satisfactory to Bank.

5. Borrower has accepted and does hereby accept the sole responsibility for the selection of his general contractor(s) if any or subcontractor(s) for the erection of the improvements herein contemplated and Bank assumes no responsibility for the completion of said improvements according to the plans and specifications and assumes no responsibility for the contract price or the due performance by the general contractor under the terms of said contract.

6. Borrower hereby consents for Bank and its agents to examine and make copies or extracts of Borrower's books and records and if and when requested to do so by Bank, to deliver to Bank copies of all contracts in connection with the construction of the improvements, and to furnish bills and receipts for any expenses for labor or materials incurred in connection therewith.

7. Representatives of the Bank, of the Permanent Lender and/or the Participating Lender if any, shall have the right to enter upon the property at any time during construction and if the work is not in conformance with the Plans and Specifications in the reasonable opinion of the Bank or its agents, the Bank shall have the right to stop the work and order its replacement and to withhold all payments from the account until the work is in satisfactory compliance with the plans and specifications. If the work is not made satisfactory within the time limit set forth in any written notice from Bank to Borrower, such failure shall constitute a default under the terms of this Agreement. In the event non-conformance with Plans and Specifications may occur, Borrower agrees to furnish Bank with Change Orders or other evidence of any material change prior to construction of said change and to request approval of said change by Bank.

8. If Bank in it sole discretion shall reasonably determine that the amount of funds remaining in said account is insufficient to complete the improvements according to the Plans and Specifications, Borrower agrees on written demand of Bank to immediately deposit with Bank such funds as Bank estimates to be necessary for the completion of the improvements and Borrower's failure to deposit such funds within 10 days after receipt of written demand by the

Figure 6-11

Bank shall be construed as an event or default hereunder. The judgement and estimate of Bank shall be final and conclusive in this respect. In the event that work should cease on said improvements for any reason whatsoever, except for reasons beyond the control of the Borrower and Contractor if any for an aggregate of fifteen (15) calendar working days, whether consecutive or not, the same shall be construed as an event of default hereunder.

9. That no materials, equipment, fixtures or any other part of said improvements shall be purchased and/or installed under a Conditional Sales Contract or Lease Agreement or any other arrangement wherein the right is reserved or accrues to any person to remove or repossess any such items or to consider them to be personal property. Borrower agrees, on demand of Bank, to execute such financing statements, security agreements or any other legal instruments necessary to insure Bank a first and superior lien on any items of property that Bank may consider to be personal property. Borrower further agrees, if requested by Bank, to furnish the names of all persons supplying any items of personal property and Bank may at its option pay such party or parties directly from the Building Loan Account.

10. Borrower will, upon request of Bank, provide a list of all contractors and subcontractors employed in connection with the construction of improvements showing the name, address and telephone numbers of each, together with a general statement of the nature of the work performed, or the materials to be supplied and the approximate dollar value of such labor or materials due each person. Bank is authorized to contact directly each of said parties to verify the information contained in said list. Bank may, at its option, pay any parties shown on said list directly from the Building Loan Account.

11. At all times during the progress of the construction, Borrower is to carry or to hire only general contractors or subcontractors or material suppliers who carry and keep in force sufficient Workmen's Compensation and Public Liability Insurance to protect the Bank and Borrower against all claims for damages sustained by any workman employed by the Borrower or by any subcontractors or material suppliers or by the public, through the acts or neglect of the builder, subcontractors or their employees on said work.

12. Borrower, at Borrowers expense, will provide insurance in the amount requested and with companies acceptable to Bank covering the improvements on the herein described property against normal hazards, including but not limited to, fire, theft, pilferage, wind-storm, flood, and vandalism, and having a Lenders Loss Payable Endorsement in favor of Bank.

13. If Borrower abandons or fails to diligently proceed with the construction of the improvements according to the Plans and Specifications, Bank shall have the right, at its option, to take over and complete the work, of construction and for that purpose to make disbursements from the Building Loan Account. Any and all contracts entered into or indebtedness incurred upon the exercise of such option may be made in the name of Borrower and for this purpose Bank is hereby irrevocably appointed attorney-in-fact, said appointment being coupled with an interest. Provided, however, that in no event shall Bank be under an obligation to complete or proceed with the construction or improvements nor shall Bank be required to expend its own funds for the completion of improvements should the funds remaining in the account be insufficient. However, Bank may, at its option, advance additional funds, and if so, Borrower agrees to pay same to Bank on demand, together with interest thereon, at the same rate as the Construction Note of Borrower.

14. Borrower shall hold Bank free and harmless from any demand, claim, or liability for the payment of any real estate commission, charge or brokerage fee in connection with the making of this loan or the refinancing thereof, and all such charges, if any, shall be paid directly by Borrower to the person or parties indicated thereto.

15. Borrower warrants and represents that he is the owner of the real property described herein and does hereby irrevocably appoint, designate, empower and authorize the Bank as agent and attorney-in-fact to execute and file for record any notices of completion, cessation of labor, or any other notice that Bank deems necessary to be filed for record to protect its interest under the provisions of this Agreement or under the Note and Deed of Trust hereinabove mentioned. The appointment of Bank as agent is expressly declared to be that of an agency coupled with an interest and as such is irrevocable.

16. Borrower hereby grants Bank the right to erect or cause to be erected Bank's sign or signs in size and location desired by Bank on the Property so long as such sign or signs do not interfere with the reasonable construction of the Improvements. Borrower will and will cause contractor if any and subcontractors to exercise due care to protect said sign or signs from damage.

17. The waiver by Bank of any breach or breaches under this Agreement or under the Deed of Trust not be deemed, nor shall the same constitute, a waiver of any subsequent breach or breaches on the part of the Borrower.

18. Bank shall have the right to commence, to appear in or to defend, any action or any proceeding purporting to affect the rights or duties of any party hereunder or the payment of any funds from or to the Building Loan Account and in connection therewith to pay from said loan account all necessary expenses including reasonable attorney's fees and for all of which the Borrowers, jointly and severally, agree to pay Bank upon demand.

Figure 6-12

19. This Agreement is made for the sole protection of Borrower and Bank, its successors and assigns, and no other person or persons shall have the right of action or rights accruing hereunder.

20. Time is of the essence of this agreement.

21. In the event of the death of the Borrower, or dissolution, bankruptcy, or assignment for benefit of creditors by the Borrower, prior to the completion of the improvements contemplated hereunder, or prior to the use or release of the entire building loan funds, the terms of this agreement shall not be affected and Bank is empowered to use and disburse said funds remaining for the purposes and upon the terms and conditions herein stated.

22. All rights and remedies of the Bank provided for herein are cumulative and shall be in addition to all other rights and remedies provided by law including a right of offset and banker's lien and the rights and remedies set forth in the Deed of Trust hereinabove referred to.

23. If this Agreement is executed by more than one person, firm, or corporation, all liabilities and obligations hereunder shall be joint and several.

EXECUTED this ____________ day of ____________, 19____.

BORROWER:

____

____

ALL CITY BANK

BY: ______________________________

NOTE: Required items in paragraph 1 to be initialled and Exhibit "A" to be signed by Borrower.

Figure 6-12A

If a completion bond is waived, the lender will not commit himself to the responsibility of the incompletion. Know the details of the agreement before you sign.

Pre-construction affidavit When all normal loan requirements are met and the final approval of the contractor's bid has been made by your lender, the final loan closing can take place with the drawing up of the mortgage documents. Before the documents are actually recorded the sworn statement from the owner and the contractor, known as the pre-construction affidavit, certifies that no work has been done or materials delivered to the site. It also certifies that no material has been worked on at another site to be used for construction of the project under the upcoming loan—which would permit a lien to be filed on this work done prior to the lender's mortgage. See Figure 6-13A and B.

The final inspection of the lot then takes place. In some states a lien law specifies that when the mortgage is recorded and the construction contracts are signed the site is immediately inspected to ascertain whether any work has been done or any material delivered. It is important for you to have done no work whatsoever on the site, as a lien can be filed prior to the mortgage. A picture is taken and a written report is completed, with copies going to the owner and the contractor. The report should indicate that an inspection was made on a certain date at a certain time of day—preferably early in the morning—to confirm without question that there has been no work performed or material delivered to the project site. The inspector will have with him his copy of the pre-construction affidavit.

Inspections and Payouts

You may not be able to ascertain in advance the number of inspections to be made by the lender, but a tentative schedule should be discussed. There is actually no limit to the number of inspections the lender may make; the program could include three or four, while some lenders feel it necessary to make ten or twelve. This depends on how much confidence he shows in you and how much supervision he thinks you need

JOINT PRE-CONSTRUCTION AFFIDAVIT

Owner

State of______________________________
County of____________________________

The undersigned being duly sworn, say that he/they_____ ____________________________________is/are fee simple owner of the premises known as______________________ which has been mortgaged to__________________for the sole purpose of securing a construction loan on said premises. The undersigned deponent states that to his/their knowledge there has been no work started, materials delivered, or labor begun either directly or indirectly up to and including this date.

Deponent further says there are no pending suits against him/them in any court or any judgments in any court remaining unpaid; and that no person has any contract for the purpose of, or claim to or against said premises with the exception of the following:___________________________
____

Further affiant saith naught.

______________________________Owner
______________________________Owner

Sworn to and subscribed before me this______day of_____19__

My commission expires_________________

____
Notary Public

Figure 6-13A

during difficult phases of the work. It pays to encourage the lender's inspector to visit your project often and at all stages of construction. In case of a slowdown or a cost overrun, your willingness to keep him abreast of developments as you notice

Contractor

State of ______________________________
County of _____________________________

The undersigned being duly sworn say that he/they_____ __________________________________ is/are construction contractor selected to construct a_____________________ _______________on the premises known as______________ _______________which has been mortgaged to____________ _______________for the sole purpose of securing a construction loan for the erection of the above structure or structures.

Deponent, as general contractor, further states that construction cost estimate has been furnished at this time to be applied to labor and materials of said structures as installed on said premises. Affiant further states that no labor or materials have been used or prepared for use, prefabricated or assembled, in anticipation for use on said premises, to this date.

Further affiant saith naught.

____
General Contractor

Sworn to and subscribed before this_____day of_______19____

My commission expires____________

____
Notary Public

Figure 6-13B

them may prevent disputes between you and your contractor.

The lender inspects the project from time to time in order to know when to pay out. Usually a written report is made. The schedule of inspections provides a check on plans and specs which can catch errors or defective work before these conditions advance very far. Sometimes an inspector is actually hired by the lender; in many cases, especially in rural

ALL CITY BANK

INSPECTION AND DISBURSEMENT SCHEDULE

(ONE TO FOUR UNITS)

For Inspection, Call:

EXHIBIT A

This Schedule refers to the Building Loan Agreement dated ______________________,

for property situated ______________________.

Before release of the FIRST FOUR INSTALLMENTS, inspections will be made by All City Bank's Inspector and disbursements will be made in accordance with his findings that the following stages of construction have been reached. It is agreed that said inspections are made for the benefit of All City Bank and not for the benefit of the Borrower and that the Borrower will not rely thereon, but will make such inspections on his own behalf as he deems advisable. Where a stage of construction requires approval by local and/or government authority, such signed approval is required prior to disbursement. All City Bank is not required to obtain receipted bills covering work done or materials furnished.

1st INSTALLMENT 15% — When foundation, underground plumbing, and floor joists in place and framing lumber on the site and upon compliance with the provisions of paragraph 4c) of the Building Loan Agreement;

OR if slab construction,

When slab poured for structure (not including garage slab), and framing lumber on the site and upon compliance with the provisions of paragraph 4c) of the Building Loan Agreement.

2nd INSTALLMENT 30% — When roof is on and rough plumbing, electrical, heating, air conditioning and framing completed.

3rd INSTALLMENT 20% — When interior plastering or drywall is completed and prime coat of paint (if frame building) or brown coat of stucco (if stucco building) on exterior.

4th INSTALLMENT 25% — When building is complete and Notice of Completion filed of record.

5th INSTALLMENT 10% — Upon compliance with the provisions of paragraphs 4b) and d) of the Building Loan Agreement.

The above disbursements refer to percentage of an amount for construction purposes in the Building Loan Account of $______________. Said construction amount shall consist of $______________, being net proceeds of the subject construction loan and $______________being borrowers funds to be deposited into the Building Loan Account prior to closing of the subject loan. These disbursements for construction purposes shall be made by check payable to:

In addition to said construction amount there shall be deposited into the Building Loan Account:

$______________ (Cash)

BORROWER:

By:______________

By:______________

Figure 6-14

	Six payouts	Five payouts	Four payouts	Three payouts	** Three payouts
Foundation & rough grading First floor laid	15%	25%	25%	50%	30%
Framing complete & subflooring Roof sheathing & chimney Window frame set	15%				
Rough plumbing, heating, wiring	10%	15%	25%		
Exterior finish Lath & Plaster or drywall	20%	20%		25%	30%
Basement floor & heating plant Interior finish, except flooring	20%	20%	25%		
All finish flooring ready for occupancy	*20%	*20%	*25%	*25%	*40%

*The 10% to be held back during the lien period will be deducted from the last payment. This period varies in accordance with state lien laws.

**Three payouts may also be made in these percentages when the home is smaller and simpler in construction (such as no basement or air conditioning).

Payout schedules
Figure 6-15

areas, the staff appraiser is the only person available for this work. Any changes proposed by you or the contractor should be thoroughly discussed with the lender. Sometimes even the smallest change can cause a chain reaction resulting in the loss of several hundred dollars. Stick to the plans and specifications so that no alteration is necessary; and make sure the lender is fully aware of the ramifications of any changes that may be needed. A lender's inspection and disbursements schedule is shown in Figure 6-14. A payout schedule comparing various numbers of payouts to percent complete is shown in Figure 6-15.

The lender is careful to see that all the contractor's bills are paid in full so there are no liens placed against the property. But you should be aware of the anticipated costs as the project progresses. Sometimes payments may be made in time periods such as bi-weekly or monthly. A contractor with considerable capital will be able to pay his labor and material bills in such a way as to have as few payouts as possible, whereas a contractor with a smaller budget may require more payouts in order to take advantage of cash discounts with material suppliers. Schedules are distributed to all parties involved. In all cases your money as the borrower will be used first. Then the proceeds of the loan totaling the full construction costs that are set aside in a special fund will be disbursed as called for in the schedule.

The payouts themselves are made by one of several systems. In one, the contractor drops in to the lender's office with his bills and the necessary waivers, and the lender issues the checks in the borrower's name with endorsements to the firms named on the checks. The contractor may arrange to meet the borrower and get his endorsement on the checks, or the borrower may accompany the contractor to the lender's office to endorse the checks. In another system, the lender provides the contractor with a contractor's affidavit similar to the one shown in Figure 6-16. The borrower may sign this form approving payment of bills attached to it. The checks are then drawn and issued to the contractor. However, under no circumstances should you sign this form in advance if you are the borrower. In the third system, the contractor obtains the borrower's signature on his bills without any specific affidavit and delivers these signed bills to the lender. If the lender

STATE OF____________________
COUNTY OF____________________ } ss

CONTRACTOR'S AFFIDAVIT

The undersigned, being first duly sworn on oath, deposes and says that he is____________________
____________________, the contractor for the____________________
work on the building situated at ____________________
owned by____________________
that the following are the names of all parties who have furnished material or labor, or both, to the undersigned for said work and of all parties having contracts or subcontracts with the undersigned for specific portions of said work or for materials entering into the construction thereof and the amount due or to become due to each, and that the items mentioned include all labor and material required to complete said work according to plans and specifications:

Name of contractor	Kind of work	Amount of contract	Amount paid to date	Bal. due or to become due
		LABOR NECESSARY TO COMPLETE		

Total $__________

The undersigned further states that there are no other contracts for said work outstanding and that there is nothing due or to become due to any person for material, labor or other work of any kind done or to be done upon or in connection with said work other than stated above: that the waivers of lien presented herewith are true, correct and genuine and are signed by the respective persons who names appear thereon; that each and every such waiver was delivered unconditionally; that said waivers were not obtained by or through any fraud, mistake or duress, and that there is no claim either legal or equitable to defeat the validity of said waivers. The undersigned further states that the amount of contract, previous payments, and balance due as shown on the reverse side hereof is a true and correct statement of the contract referred to herein, and that all labor is paid in full to date.

That the undersigned makes this affidavit for the purpose of procuring from the owner and__________
____________________ a payment upon contract for said labor or material or both.

Signed this________day of________________19______

____

____________________ Address

____________________ Phone

SUBSCRIBED and SWORN to before me this
__________day of__________A.D. 19__________

____
Notary Public

Contractors affidavit
Figure 6-16

concurs with them, he pays the contractor immediately. This third system is used more in smaller communities. The payouts are usually made by the lender directly to the contractor or supplier, and a waiver of lien is obtained from each one when the final bill is paid. This protects you and the lender from any mechanic's liens or lawsuits which may arise because of unpaid bills. Many times the lender requires the owner-borrower to physically participate in the payouts to establish positive proof that they were disbursed for labor and materials.

Payouts should always lag behind the equity in your project, and under no circumstances does the lender make the payouts before a particular stage of the work is finished as called for in the schedule. Although you may not actually make the payments, you should observe this part of the loan program. Your lender will be pleased by your observation on this point. Sometimes good service among the trades can be increased by tactfully discussing with your lender when a sub or a particular supplier may expect his money. You should keep a running estimate during construction to see that the equity in the project exceeds the payouts to date by a fairly good margin.

To assure the availability of funds set aside for the entire project, a lender usually holds out about 10% from the progress payments because some work such as plumbing, wiring, heating and equipment as well as plastering often needs adjustment even though the job is completed. This hold-out is the most dependable way of assuring a final and satisfactory installation. It is expected among contractors and is an established and a customary procedure. It has nothing to do with integrity or workmanship.

You can expect the lender to see that there is no actual moving into the property until the project is finished. The incomplete project will not be considered finished and ready for occupation until all stages of construction are over and inspected. Some owners lose interest in the last stage; but for the most part, only a small amount of encouragement from the appraiser or the lender's representative sees the whole job through.

If the above program has been carried out with no misunderstandings between the principal parties, then waiving all liens will assure that no mechanics lien exists at this late stage of completion.

Mechanic's Liens

You can assure yourself of no mechanic's liens by paying

all bills for construction and improvement, labor and materials. No such lien can exist if bills are paid.

Mechanic's liens are debts holding precedence over all other types of liens. Mechanic's lien laws are the result of state legislation to protect the laborer and the materials supplier in the event of the owner's or builder's insolvency. Consult the laws in the states where you own property. Lenders throughout the country are constantly in touch with their attorneys to protect their savers' deposits, and since they are in the real estate loan business, they can be relied on to help you avoid a lien that could prevent or delay repayment of your construction loan.

If there were no lender involved in the project whatsoever, any claim for payment covering overdue funds for contract work would be a burden only to the owner of the property and its improvements. As indicated before concerning the program for payouts, lenders make a special effort to obtain waivers of lien for contractors in order to assure that those involved in the project have been paid according to your contract. If a lender advances no money he does not look out for an owner's interest because the assets of his depositors are not in jeopardy. But any lawsuits relating to an owner's debts would affect any mortgage. The lender makes an investment in real property; and a recorded lien, even though not justified, may cause expensive legal proceedings and possible work stoppages the lender tries to avoid if he can.

Who Has the Right to a Mechanic's Lien

Those who contract with you specifically to be paid for services to the construction project are contractors because you and the contractor hold a legal agreement which has as its main purpose the improvement of real estate properties. These persons are contractors if they furnish labor or materials in anticipation of constructing a building—or if they repair or alter it, improve the land or fill, sod or do landscape work on it. Contractors also include those who represent themselves as engineers, architects, or superintendents. In most states a mechanic's lien arises—but may not actually exist—at the moment your contract with any one of these persons is consummated through either an express or implied initial agreement. The debt must be proven to exist at the moment of the agreement; so if the time period starts with the agreement or contract and runs through the period of construction, it must also have a legal expiration date. This period of time during which the contract is still valid after the work is completed is usually four months. This protects innocent third parties who might purchase the property or place a mortgage on it subsequent to the completion of the work. Although this period may be extended by mutual agreement between the owner and the contractor, the lien does not necessarily have to be recorded to be effective during the four months. Because mechanic's liens have priority over other liens, they are automatic when actual written notice has been made or recorded in the customary manner.

A mechanic's lien is evident when it is obvious that material is on a building site, whether it is recorded or not. It is for this reason that the lender gathers all the information he can at the time of recording the mortgage. If there is no evidence of a lien on the record, no material delivered to the premises and no prevailing questions or doubts of a contract nature—oral or otherwise—then the construction loan will be recorded.

Because mechanic's liens are automatic when actually recorded or when other official notice is made, there must be a period of time specified after the materials have been delivered, or the labor provided, in which notice must be served on the owner. This is a separate and distinct requirement which should not be confused with the time in which the contract is valid. State statutes vary on the notice period. The most common is sixty days. The pre-construction affidavit precludes the sixty day notification period.

An additional safety provision may be included in the building construction contract pertaining to the subcontractors' liens, as their rights are included in this contract. You can negotiate this with the contractor. Some state statutes demand that you record this type of contract before subs begin work; in this way the agreement may avoid their rights to lien.

7

Home Improvement Financing

In remodeling, as in all the other phases of construction, good building methods must be used to hold down costs without creating design deficiencies that may increase maintenance, decrease livability, or lower the resale value of the property. And the remodeler must be a specialist not only in the building trade but also in the financing of his ventures. As an improvement specialist you should be aware of this specialty demand and how it applies to your selling an improvement package. You may have to promote the financing arrangements yourself. Look into local government remodeling programs. The planning branch of the local building department lists improvement loans available to your customers, such as HUD development programs supported by city governments. If your customers cannot get the money for your remodeling proposal, you will be able to assist them with this information.

Large building firms are able to take advantage of methods used to reduce costs in spite of inflationary conditions. These builders often build hundreds of units a year. Because of their volume needs, they can buy materials directly from the manufacturer and can develop their building sites from large sections of land. Much of the work, such as roofing, gypsum board interior finish and painting, is done by subcontractors. These crews are specialists, each one becoming proficient in its own phase of the work. Central shops are often established where all material is cut to length and often preassembled before being trucked to the site. These methods reduce the cost of a single house in larger projects. But most of these methods cannot be applied to the individual remodeling job, nor even to the individual house being built by the owner.

The home improvement loan has for many years been an important part of savings institutions' programs. There are many reasons why savings associations offer this type of credit to the public. In recent years the most important reason has been the strong demand. As far as most lenders are concerned, the remodeling operation, although not a big profit maker, is economically sound. Lenders in larger communities consider that the enhancement and conservation of home values in the community provide them better security for mortgage lending. A home improvement loan is a very safe loan. It is considered liquid because of its short maturity and a ready marketability. As a sideline to granting regular mortgages, the remodeling loan brings more people into contact with the lender who can be developed into savers and mortgage loan borrowers

The open-end mortgage feature in many existing loans meets part of the demand for home improvement credit, the security portion of the real estate. However, the open-end device can never meet the needs of the general homeowning public. The unsecured property improvement loan serves an entirely different group of people. The improvement contractor should be geared to this type of home improvement financing, due to the flexibility of terms or the selection of lenders.

The *improvement contractor* is in a position to promote his services by virtue of his free-lance ability to solicit clients for an addition, expansion, or modernization. Naturally, he determines whether or not the prospect he has contacted for an improvement has the type of mortgage on his property that can be used to merely advance the mortgage. Although the work might be contracted with the owner's existing open end mortgage, if he has one, the owner must still take the initiative

Your application for Dealer-Contractor Approval has been passed by the Association's Loan Committee. We wish to welcome you to the Association and look forward to a satisfactory relationship.

We would like to take this opportunity to outline procedures in the purchasing of your FHA Title 1 paper upon completion of the job.

Submit a complete set of papers as follows:

1. Credit Application and Note.

2. Completion Certificate.

3. Copy of The Contract.

Please take particular care that date of note and date of completion is the same.

The first payment cannot exceed 60 days from date on the note. No deals can be purchased until after six full days from date of the commitment.

Our Home Improvement Credit Department is ready to render fast service to you at 001-0001

Very Truly Yours,

____

Loan Officer
Improvement Loan Department

A Letter of Dealer-Contractor Approval
Figure 7-1

with his lender in any proposal to modernize. This is sales resistance, as far as the contractor is concerned, if he must wait for an owner to negotiate with his mortgagee.

An unsecured property improvement loan is a personal loan rather than a real estate loan as such. It is nothing more than consumer credit. Most savings associations will make unsecured property improvement loans directly with an improvement contractor. A lender establishes a business relationship with this type of contractor in advance. It is up to the contractor to find his jobs on the basis of estimates, proper design, plans and specifications and, best of all, no construction delay once an agreement is signed and approved. There is no delay because the contractor does not have to wait for the loan to be disbursed from the lender to the homeowner to the contractor. In addition, if the contractor has operating capital for use in his bargain hunting for material, no delay will result in the construction starts. He has a contract with the owner on which to base the cost of the project's material. Cash on hand allows the remodeler to pick up timely cash discounts from suppliers.

Lenders often prefer to buy the completed loans from the contractor after they are filled out and signed by the consumer. This, of course, depends upon the customary system of each lender. If he has a list of approved contractors he knows and intends to continue to do business with, an implied contract is established and the improvement contractor should have no fear of obtaining his money from the lender upon completing the improvement. The unsecured property improvement loan is a key to successful improvement deals.

Figure 7-1 shows a contractor's formal approval letter from a lender. In some cases a financial statement may also be required although in some areas and smaller communities this requirement may be waived. A typical financial statement is shown in Figure 7-2. The improvement contractor can promote his deals faster and more smoothly as his own agent-contractor acting as a representative, or at least a go-between, in arrangements with owners and lenders. Try not to rely on an improvement contract that is subject to an open end mortgage handled by the owner. The credit application for the improvement loans (Figures 7-3, 7-4, 7-5, 7-6) is submitted

FEDERAL SAVINGS AND LOAN
Financial Statement

Date:________________

Name of Company______________________________
Location____________City____________State__________

Company Assets

Accounts receivable	$ ____________
Cash	$ ____________
Equipment	$ ____________
Autos or Trucks	$ ____________
Other assets	$ ____________
Total	$ ____________

Company Liabilities

Accounts payable	$ ____________
Notes payable	$ ____________
Taxes withheld	$ ____________
Liens on all trucks, equipment, etc.	$ ____________
Total	$ ____________
Net Worth	$ ____________

By:____________________ Title:________________

Dealer-Contractor Improvement Application
Figure 7-2

under the provisions of Title 1, National Housing Act. The HUD form number is FH-1 (2-78). Since this improvement loan is not a real estate-secured loan, as a mortgage instrument would be, a Title 1 method is preferred. The reverse side, Figures 7-5 and 7-6, is the certificate provision of the

U. S. DEPARTMENT OF HOUSING AND URBAN DEVELOPMENT
HOUSING- FEDERAL HOUSING COMMISSIONER

CREDIT APPLICATION FOR PROPERTY IMPROVEMENT LOAN

This application is submitted to obtain credit under the provisions of Title I of the National Housing Act
(PLEASE ANSWER ALL QUESTIONS)

FH-1 (2-78)

TO: Lending Institution which will provide the funds: | DATE:

1. Do you have any past due obligations owed to or insured by any agency of the Federal Government? *(If the answer is "Yes", you are not eligible to apply for an FHA Title I Loan until the existing debt has been brought current.)* *Check Block* ☐ Yes ☐ No

2. Have you any other application for an FHA Title I Improvement Loan pending at this time? ☐ Yes ☐ No *(If "Yes", with whom - name and address.)*

3. I hereby apply for a loan of $ ________ (Net) to be repaid in ________ Months.

4. APPLICANT (S)

Applicant Name: | Sex | Age | Number of Dependents

Marital Status: ☐ Married ☐ Unmarried (including Single, Divorced, Widowed) ☐ Separated

(Check Applicable Box): (1) ☐ American Indian or Alaska Native (2) ☐ Asian or Pacific Islander (3) ☐ Black, not of Hispanic origin (4) ☐ Hispanic (5) ☐ White, not of Hispanic origin (6) ☐ Other (specify)

Co-Applicant Name (If any): | Sex | Age | Home Phone:

Marital Status: ☐ Married ☐ Unmarried (including Single, Divorced, Widowed) ☐ Separated

(Check Applicable Box): (1) ☐ American Indian or Alaska Native (2) ☐ Asian or Pacific Islander (3) ☐ Black, not of Hispanic origin (4) ☐ Hispanic (5) ☐ White, not of Hispanic origin (6) ☐ Other (specify)

The information concerning minority group categories, sex, marital status, and age is required for statistical purposes so the Department may determine the degree to which its programs are being utilized by minority families and for other evaluation studies.

Credit application
Figure 7-3

dealer-contractor or the salesman representing the contractor.

The unsecured loan application is partially filled in by the contractor as to the actual amount needed, the term of the loan, and other contract details. The prospective borrower must complete the application and return it, signed, to the contractor. Credit references and former employers must be revealed by the consumer. This process may be speeded up by the association; it can obtain most of the information by telephone, requesting the homeowner to drop in at his convenience for signing. Even better, the improvement contractor can sometimes drop in the next day for the signatures.

You will be notified by the lender when your application has been fully approved. As a contractor, you will find it expedient to deliver a promissory note payable to you after execution by the borrower. The lender then makes the note a part of the loan file. When the work is satisfactorily completed, evidence in the form of an improvement certificate signed by the borrower will then be made available to the lender. The borrower signs the certification if he is satisfied that the workmanship and materials are just as expected of a

Address (Street, City, State and ZIP Code):	How Long	Name and Address of nearest relative not living with you	Relationship
Previous Address (Street, City, State and ZIP Code):	How Long		

5. EMPLOYMENT AND SALARIES: *(If applicant is self-employed, submit current financial statement.)*

Employer Name and Business Address:	Type of work or Position :	No. of Years:	Business Phone	Salary (Week/Month)
				$ per
Previous Employer Name and Business Address:				
Co-Applicant's Employer Name and Business Address:	Type of work or Position:	No. of Years:	Business Phone	Salary (Week/Month)
				$ per
Other Income-Source *(Note: Income from alimony, child support, or separate maintenance income need not be shown unless you will rely upon it as a basis for undertaking or repaying this loan.)*				Amount (Week/Month) $ per

6. BANK ACCOUNT:

☐ Yes ☐ No ☐ Checking ☐ Savings	Name and Address of Bank or Branch:

7. CREDIT ACCOUNTS: (Give name and address of finance companies or stores which have extended credit and which you have paid in full.)

a.	b.
c.	d

8. DEBTS: List all fixed obligations, installment accounts, FHA loans, and debts to banks, finance companies, and Government agencies. *(If more space is needed, list all additional debts on attached sheets.)*

FHA Ins. Yes	FHA Ins. No	To Whom Indebted (Name) Mortgage/Contract	City and State	Date Incurred	Original Amount	Present Balance	Monthly Payments	Amount Past Due
					$	$	$	$
					$	$	$	$
					$	$	$	$
					$	$	$	$
					$	$	$	$
AUTO		Lien Holder:		Year and Make:		$	$	$
		Lien Holder:		Year and Make:		$	$	$

Previous Editions are Obsolete FH-1 (2-78)

Figure 7-4

professional craftsman. The lending institution then buys the note as arranged between the parties, and the net proceeds of the loan are then disbursed directly to the improvement contractor. In some cases the lender may service the construction improvements to the degree that actual inspections are made before final purchase of the note. An improvement completion certificate, called HUD Form FH-2, is shown in Figure 7-7.

-2-

9. PROPERTY TO BE IMPROVED:				
Is this a new residential structure, has it been completed and occupied for 90 days or longer ? ☐ Yes ☐ No				
Address (Number, Street, City, County, State and ZIP Code):			Type - Home, Apt., Store, Farm, etc. , (If Apt., Number of Units):	Date of Purchase:
Census Tract:			Year Built:	
FILL IN ONE: Is Owned By:	Name of Title Holder:		Date of Mortgage:	Price Paid: $
Is being bought on Installment Contract By:	Name of Purchaser:	Name and Address of Title Holder:		Price Paid: $
Is Leased By:	Name of Lessee:			Date Lease Expires:
Name of Landlord :		Address:		Rent Per Month: $

10. PROCEEDS OF THIS LOAN WILL BE USED TO IMPROVE THE DESCRIBED PROPERTY AS FOLLOWS:		
Describe each improvement planned	Name and Address of Contractor/Dealer	Estimated Cost
		$
		$
		$
		$
		$

Figure 7-5

WARNING
Any person who knowingly makes a false statement or a misrepresentation in this application or causes such a false statement or misrepresentation to be made shall be subject to a fine of not more than $5000 or by imprisonment for not more than 2 years, or both, under provisions of the United States Criminal Code.

IMPORTANT · APPLICANT READ BEFORE SIGNING

The selection of a Contractor or Dealer, acceptance of materials used, and work performed is your responsibility. Neither the HUD-FHA nor the Financial Institution guarantees the material or workmanship or inspects the work performed.
I (We) certify that the above statements are true, accurate, and complete to the best of my (our) knowledge and belief. This application shall remain the property of the Lending Institution to which submitted for the purpose of obtaining a loan.
I (We) hereby consent to and authorize the Lending Institution or the HUD-FHA, after the giving of reasonable notice, to enter the improved property for the purpose of determining that the improvements specified in this application have been completed.

Name ______________________ (LS) Name ______________________ (LS)
(Applicant) (Co-Applicant)

NOTE TO SALESMAN: If proceeds will be disbursed to the Contractor/Dealer, the person(s) selling the above described improvements must sign the following certification.

I (We) certify that: 1) I (We) am (are) the person(s) who sold the job. 2) The Contract contains the whole agreement with the borrower. 3) The borrower has not been given or promised a cash payment or rebate nor has it been represented to the borrower that he or she will receive a cash bonus or commission on future sales as an inducement for the consummation of this transaction; that the improvements have not been misrepresented; no promises impossible of attainment; no encouragement of trial purchase; no promise that the improvements will be used as a model for advertising or other demonstration purposes; and no offer of debt consolidation.

(LS) Name ______________________
(My true name and signature are as shown above)

If application is prepared by one other than the applicant, the person preparing the application must sign below.
I (We) certify that the statements made herein are based upon information given to me(us) by the borrower(s) and are accurate to the best of my (our) knowledge and belief.

Prepared by: ______________________ Address: ______________________

Representing: ______________________
(Name of Dealer/Contractor)

(Reserved for use of Lending Institution):

FH-1 (2-78) Previous Editions Obsolete

US GOVERNMENT PRINTING OFFICE: 1978-789-099/97

Figure 7-6

U.S. DEPARTMENT OF HOUSING AND URBAN DEVELOPMENT
FEDERAL HOUSING ADMINISTRATION

COMPLETION CERTIFICATE FOR PROPERTY IMPROVEMENT LOAN
(UNDER FHA TITLE I)
WORK DONE OR MATERIALS DELIVERED

TO: (Financial Institution)	Address

In accordance with my (our) credit application dated ______________________, for a loan pursuant to the provisions of Title I of the National Housing Act:

I (We) certify that I (we) have not been given or promised a cash payment or rebate nor has it been represented to me (us) that I (we) will receive a cash bonus or commission on future sales as an inducement for the consummation of this transaction.

I (We) understand that the selection of the dealer and the acceptance of the materials used and the work performed is my (our) responsibility and that neither the FHA nor the financial institution guarantees the material or workmanship or inspects the work performed.

CHECK HERE IF LOAN IS TO PAY FOR COST OF MATERIALS AND INSTALLATIONS.

☐ I (We) hereby certify that all articles and materials have been furnished and installed and the work satisfactorily completed on premises indicated in my (our) credit application.

CHECK HERE IF LOAN COVERS ONLY THE PURCHASE OF MATERIALS

☐ I (We) hereby acknowledge receipt in satisfactory condition of the materials described in my (our) credit application.

NOTICE TO BORROWER	DO NOT SIGN THIS CERTIFICATE UNTIL THE DEALER HAS COMPLETED THE WORK AND/OR DELIVERED THE MATERIALS IN ACCORDANCE WITH THE TERMS OF YOUR CONTRACT OR SALES AGREEMENT.	SIGNATURES OF BORROWERS	DATE SIGNED
		(Read Before Signing)	
		(Read Before Signing)	

For the purpose of inducing the payment of proceeds of this loan and the insurance thereof by the FHA the undersigned certifies and warrants that:

(1) The above work or materials constitute the entire consideration for which this loan is made.
(2) A copy of the contract or sales agreement has been delivered to the borrower and the above financial institution.
(3) This contract contains the whole agreement with the borrower.
(4) As an inducement for the consummation of this transaction, the borrower has not been given or promised a cash payment or rebate nor has it been represented to the borrower that he will receive a cash bonus or commission on future sales.
(5) The work has been satisfactorily completed or materials delivered.
(6) The above certificate was signed by the borrower after such completion or delivery.
(7) The signatures hereon and on the note are genuine.
(8) All bills for labor or materials have been or will be paid within 60 days and that the improvements had not been misrepresented to the borrower.

If any of the above representations prove incorrect, the undersigned agrees to promptly repurchase the note from the financial institution or from the FHA as the case may be.

DEALER SIGN HERE	Name of Dealer	Date
	Signature	Title

WARNING

Any person who knowingly makes a false statement or a misrepresentation in this certificate shall be subject to a fine of not more than $5,000 or to imprisonment for not more than 2 years, or both, under provisions of the United States Criminal Code.

3. DEALER/CONTRACTOR COPY

Improvement completion certificate
Figure 7-7

The real difference between the open end mortgage advance and an improvement loan as such, is the degree of control that the improvement contractor maintains during the course of the loan process. The builder has no control over the borrower's open end mortgage advance, and is obligated (and expected) to make the estimate, complete the remodeling satisfactorily, and make available the completion certificate at the termination of the contract.

Institutions dealing in home improvement loans as a specialty are particularly strict as to the actual use of the money. The improvement lender will not be dictatorial, but you must give him an assurance that the loan is for home improvement only (contrary to the refinance loan). Improvement finance proceeds are limited to the following categories: remodeling, expansion, alteration, conversion, restoration and conservation.

To qualify for an improvement loan, the plans and specs must be for one to four family residences that are not less than 10 years old since original construction.

Cost Reduction and Remodeling

The design or layout of the remodeling project is an important factor that the lender may be particularly interested in because he prefers to minimize cost. Select the plan that permits the use of standard length materials for joists and rafters. When planning a project, obtain advice from an architect, engineer, or other reliable authority as to the best means to accomplish material and labor savings.

How the funds are obtained, whether out of the owner's pocket or through borrowing, is actually immaterial when it comes to reducing costs. If a lender is involved, he will be interested in studying the room arrangements to see that plumbing and heating lines will have short runs. A rectangular plan without variation is usually the lowest in cost. The type of foundation to be used, such as slab, crawl space, or basement, is an important consideration for the lender. Base the foundation selection on climatic conditions and the needs of the family. While space in a basement is not so desirable as in areas above grade, its cubic foot cost is a great deal cheaper. A slab addition, including utility and storage as well as recreation space, may be as costly as one with a full basement.

A flat or low-pitched roof is often lower in cost when ceiling joists and the roof rafters can be combined. Construction is simpler and less material is used than in most pitched roofs. Design is the major factor in this type of roof, so overall appearance will be pleasing and in keeping with the original structure. Pitched roofs should have simple lines for lower costs. Variations in the roof add to the framing costs; however, slight changes are desirable to relieve the monotony of a straight roofline. If dormers are needed for an expandable attic, they should be included in your construction program. Dormers, though, can add considerably to the costs and should be justified.

The choice of materials is important to remodeling cost savings. The use of concrete blocks for foundation walls should be balanced against the use of poured concrete or other masonry materials. These costs may vary by areas. Precast blocks for chimneys may be considered if they are available. These blocks are made to take flue linings of varied sizes and are laid up more rapidly than brick. The use of prefabricated lightweight chimneys that require no masonry may also save money.

Dimension material varies in cost as to its species. Use the better grades for joists and rafters and lower grades for the studs. Do not use grades which involve cutting and selection procedures, or the savings will be dissipated by increased labor costs. Do not use better grades of lumber than are actually needed.

The cost of exterior finish may vary considerably. For example, while more pleasing in appearance, wide, thick-butt siding may cost 50% more than the narrower types. Depending upon the design of the existing structure, the cost of the wider siding may be justified. Also, rust-resistant nails are well worth the additional costs for siding. Species of siding lumber should be considered not only on the basis of original cost but also on their paintability.

The choice of interior covering should be considered. While drywall construction may be lower in cost per square foot, it requires painting before it can be considered complete—whereas plaster walls do not require immediate painting.

These costs vary by areas, depending on availability. A home improvement lender can probably be of some help in directing you to suppliers he may be acquainted with unless you are fully equipped and do not need such assistance.

The choice of flooring, trim, and other interior finishes should be carefully studied. Lower cost trim species are ordinarily painted, while oak and birch may be given a coat of sealer only. Use standard moldings and stock window sizes to decrease the cost. Use standard sizes for cabinets when possible. Custom built material of this type is usually much higher in cost.

Horizontal assembly of full wall sections, even to application of diagonal bracing and sheathing, is used by many contractors to advantage. These sections are raised in place and fastened to the floor system. Where gypsum board drywall is used, many contractors use the horizontal method of application. This way, the taped joints are brought below eye level for ease and speed of taping and room-size sheets may be used to cut down on taping labor and material cost. Vertical joints may be made at windows or the door openings only. During construction and remodeling, the advantages of a simple plan and the selection of an uncomplicated roof become obvious. There is less waste by cutting down on joists and rafters, and erection is more rapid than on a more intricate, highly-engineered structure.

Cost reduction takes talent, so does the acquiring of operating capital. If capital is readily available for timely discounts, the result is capitalization.

Your selling techniques are your key to an initial improvement contract while the job itself must be professional. A word of mouth reputation will supply plenty of work for the improvement contractor. And if he advertises locally that his firm specializes in improving the living conditions of his customers — enhancing their lives, giving them peace of mind, and so on — these advertising practices will bring him the business he needs to build up cash on hand. Merchandising procedures are so important that a contractor must consider operating capital for cash material purchases second only to the improvement contract itself. Quickly obtaining funds after finishing an improvement contract is also vital because you may be tied up with other money frozen at the time you made the contract. There are other sources of money besides what you have in your own account, as will be shown in the following section. The sooner the project can start after the contract is signed the better for all.

The Contractor-Financier

If a builder's regular money source is not available for an improvement loan and if his prospect wants to begin remodeling immediately, the builder can find out how much equity his prospect has accumulated in his existing property. When he is convinced of his prospect's integrity, his net worth, credit standing and other financial particulars, he can then propose a junior lien with himself as the lender, contractor, creditor and beneficiary. With existing operating capital and connections with various suppliers, all the builder needs to do is determine the amount of the contract price, the face amount of the second trust deed or mortgage, the interest rate, the rate of pay off, the escrow (or lawyer's fee), the cost for a policy of title insurance or abstract of title, and any specific terms such as an acceleration clause and prepayment penalty.

For example, the property to be improved is worth about $80,400. The improvement and remodeling along with contractor's profit, will add up to about $20,000. The equity is $42,400. The builder knows that junior lien investors use the rule of thumb of no more risk than half of the equity should be taken. The contractor will, however, charge 10% interest on a rate of pay-off of 2% with monthly installments required for a two-year period. At the end of this period, all of the balance will be due and payable. These terms are fairly standard among private investors, but are different than the terms of an institutional lender. An institutional lender stretches the repayment period out for several years while charging the annual percentage rate covering charges for making the loan. A rate of pay-off at 2% is $400 if junior lien is $20,000.

The owner would rather pay the contractor for work in this way than advertise for a private lender, who would be a total stranger. If this transaction takes place in a state permitting escrow services, the details of this contract and all necessary realty services will be provided in escrow. In other states the details are completed in a lawyer's office, if required. The owner plans to completely refinance his existing first loan in two years, or possibly sell his home when the builder (as the mortgagee) declares the note due and payable. There is nothing mandatory in the terms the owner-borrower and the contractor-financier must follow, unless it is the state regulations. There is no broker either, and no middleman involved in such a transaction. There are no commissions, nor are there any legal dictates regarding loan fees where two parties enter into a real estate secured loan identified as a junior lien. The builder is not serving as a loan broker, nor is the owner-borrower obtaining cash proceeds on a loan.

Since there is no cash advance from any lender involved, an improvement contractor can state his terms. With a price of $20,000 covering his labor, material and ordinary profit, the builder can collect over $3,400 in interest that the owner-borrower over the two year period would ordinarily pay an outside lender. Since this is a secured loan, however, there are administrative costs that a normal second mortgage or trust deed would require. The owner should bear the out-of-pocket costs for the escrow and other clerical and recording fees.

A schedule for this direct reduction loan would look like this, as shown in Appendix VII:

Loan $20,000 Rate 10% Payment $400 Term 2 years Periods 24

Payment Number	Net Interest	Principal Payment	Balance of the loan at each monthly installment
1	$166.66	$233.34	$19,766.66
2	$164.72	$235.28	$19,531.38

But instead of this time consuming process, the builder can determine the owner's balloon, or final lump-sum payment, by using the rate of pay-off formula in Appendix IV. He can then use a 2% rate of pay-off to conform with the $400 payment required:

24 ($923-$400) x ½ (2 x .10) plus 1.01
24 ($523) x ½ (.20) plus 1.01
$12,552 x (.10) plus 1.01
$12,552 x 1.11 equals $13,932.72

The amount due and payable after 24 months is, therefore, a $13,932.72 balloon. The interest collected by the improvement contractor totals approximately $3,429.

The builder can cash out the second trust deed or mortgage to a discounter if he needs operating funds, and can write the contract to attract investors. For this, he adds an acceleration clause and a prepayment penalty, which are standard in the lending world for hard money junior liens. But in a two-party agreement it credits the cost of construction work. The builder must write the terms to make the arrangement as attractive as possible to the owner. Although the second mortgage or trust deed is recorded as such, this is only added security for payment of the construction work as far as the builder is concerned. If he should *sell* the note, some loss would surely occur no matter how well it was written. If an opportunity to *borrow* on the note occurs later, he would be in a much more secure position. He still holds a very secure income on the note but probably does not need a prepayment clause to present to a hypothecator.

Therefore, the junior lien should contain an acceleration clause indicating a legal declaration by the holder of the note, that all monies are now due and payable, should the property owner default in his payments. As for the importance of a prepayment clause, the builder would prefer a full or partial prepayment. The sooner he gets paid off the better. A possible investor buying the lien whose money is working for interest value would have a different philosophy than a remodeler furnishing labor, materials and professional know-how. A prepayment clause is a penalty for paying off early; therefore, an investor may recover interest lost with such a clause included in the terms.

The owner and the improvement contractor, therefore, have a deal calling for specific terms which, if not honored, may result in a notice of default and foreclosure on the property, in accordance with state laws in which the property is located. This is no different than a bank, a savings institution, or a private investor-lender who requires this clause on a loan protected by the value of the entire property.

Private and Institutional Improvement Loans

Sometimes a builder or improvement contractor can convince an institutional lender to take over a second mortgage that has been taken back in lieu of cash on a sale or on an improvement deal. If there are already established business relations with the lender, it is good business to arrange for the institution to give credit to the builder-contractor on an account as an added security for the institution, rather than having it pay the builder in cash for the junior lien. When the second mortgage is finally paid off, the account is released to the builder. This avoids the problem of collections that the contractor has to contend with otherwise. Also, an owner will be more alert to the necessary installments required by an institution than he would be if installments were made to an individual. The institution has gained for two reasons: the interest rate is high, and the lender has pleased a good customer by purchasing the junior lien. Even more ideal, this could be the institution which also holds the first mortgage on the house but cannot for individual reasons advance money on an open-end mortgage on terms satisfactory to the owner. In this case, there would be little question about the institution holding the second mortgage against this particular property, since it already is the mortgagee on the first loan. It is also an advantage to the holder of the first mortgage to be in a position to routinely look into payments on a junior lien. Any default on the second loan very often spells trouble for the first loan. Although the first mortgage holder was not a party to the credit arrangement made for the purpose of remodeling the home, anything that creates a possible drain on the income of a mortgagor may just possibly affect the conduct of the payments on the first loan. Therefore the institution cannot be indifferent.

Aside from refinance or open-end advance procedures, a cash amount contributed by the owner as part of the project cost is always a possibility. The balance then needed could be financed through a consumer loan. Or a private investor may be contacted for hard money through newspaper want ads where more flexibility is shown than will be experienced from institutions. In this case the owner merely obtains cash secured by a 2nd trust deed. Possibly the improvement contractor wants to carry his own "paper" in the form of a second trust deed or mortgage. Depending upon circumstances, he may decide to sell the paper to a savings bank willing to credit a special account to the contractor. In this case, the contractor must have operating capital for the job at hand while drawing on the account as needed.

Or, given the same junior lien circumstances, the contractor may have a compelling reason to discount the paper. A loss would probably occur without an appropriate compensation for the difference. In the builder's case, an amount representing the interest potential, or $3,429 in our example, would be sacrificed unless it is added to the contract price. Also, the overall profit on the remodeling job would be lost if sold to a discounter. To add the interest and the profit together and then increase the contract price in order to obtain cash on a discount is the same as working for love to beat the competition. The money-making potential for an improvement contractor lies in your retaining a junior lien with no outside involvement, while using your own operating capital.

In advance of contracting for the improvement on the property, financing the job for an owner on a junior lien means preparing for an offset in the contract price to the owner. A disclosure should be made to the owner that he must bear this extra amount by providing such terms in the contract. A provision should also be made that installments may be collected by a private investor in the event the contractor does not retain the paper. This could be a discounter or a hypothecator, in which case these individuals are not a part of the improvement contract but merely holders of the note. A *completion bond* should be furnished, or a builder's control service be utilized on the owner's behalf.

Maybe the contractor wants to merely profit from the construction work itself, rather than retaining any second loans for investment. Then he should assign the note rather than discount it. Operating capital can be obtained on the collateral he holds to about the same amount as would be received if he discounted a $20,000 second loan. This is hypothecation.

A hypothecator will loan up to 70%— or 75% in some instances, depending upon the prevailing market. The rate, or in this case, the LVR based on the value of the "paper," may depend significantly on how well it is written. As a protection to both lender and homeowner, this proposed mortgage should be an amount which, when added to any outstanding indebtedness related to the property being improved, would keep the total indebtedness against the property within the limits of safety. Thus a homeowner should not be encouraged to undertake a home improvement debt burden that would be excessive in relation to the property value. This does not mean that you must also be an appraiser. But common sense should be used in ascertaining the loan to value ratio as well as other burdens the homeowner may be carrying, including the proposed junior lien. A hypothecator will check out these very points before he makes a commitment.

The owner should be aware of the terms of the junior mortgage, such as its acceleration clause, prepayment penalty, and others, that would prevail with a collateral lender just as they would if the paper were transferred to a discounter. In no case would the contractor be servicing the mortgage while also busily involved in the construction work, unless he retains the instrument himself as his personal investment. The servicing of a hypothecated second loan amounts to collecting monthly installments from the property owner (in our example $400) and returning to the borrower on the loan (in this case, the builder) the portion of the installment which represents the principal and interest on one-quarter of the remaining principle balance. In our example, the contractor has obtained $15,000 in proceeds (75% x $20,000) as operating capital. He still retains actual ownership and the prerogatives on the mortgage. It is pledged as security only. The original rate of pay-off, the time period, the rate of interest and all terms and conditions as first written, do not change.

Fees

An oversimplification has been used in our figures in order to illustrate as clearly as possible the differences between (a) refinance or open-end transactions, (b) home improvement loans or consumer credit for improvements, (c) private hard money loans from outside investors, (d) credit advances by institution to contractor, (e) discounting the loan for operating cash, and (f) borrowing on the collateral value for operating cash. This simplification is a matter of routine application and administrative procedures on unsecured as well as home improvement loans. The institution has an application fee, an originating or closing charge, reasonable and customary amounts for recording fees and charges for recording taxes or other charges incidental to recording new loans. It also charges for credit reports, title examinations, title insurance and such other reasonable or customary fees and charges including a possible appraisal fee.

These services are routine requirements for an institution making a second loan with property as security. But the items *not* in the unsecured improvement or consumer loan are a recording fee, title examination fee, a title insurance policy and an appraisal fee.

An investor advancing a hard money loan on property will request his escrow officer or the title company to list the services that must be a part of his secured loan. The second trust deed or mortgage usually requires the same routine services from private lenders that are required for any real estate loan secured by the value of the property, with the possible exception of the appraisal fee.

By far the lowest fees to be paid by the property owner are the service fees required by escrow service, or legal service, in a credit advance by the contractor. The main fees are in the recording of the junior lien, and for a title policy or abstract of title. Other incidentals should be waived in the owner-contractor relationship. There may be further requirements on the part of the hypothecator in the event that a collateral loan is made, however.

In assignment or discount sale of a second loan for improvement, a completion bond or some form of completion insurance is a must. Where private money is involved in the financing, a junior lien is secured by the value of the entire property and equity. A stoppage of the project, for whatever reason, will not affect any mortgage obligation on the remainder of the property.

8

Trade-In Financing

In a locally slow or a generally sluggish housing market, how should you handle your marketing problem in the absence of qualified buyer? You may need to consider another type of financing, the *trade-in.* Your project is ready to be sold and the lenders' short-term construction loan must soon be paid off, but there are few offers to buy.

Since construction lenders derive their profit from a short term repayment, both you and the lender can consider as money lost any delay in construction, marketing or the non-availability of a permanent loan beyond the borrowing period. For instance, the average interest rate on a 10% six-month construction loan with a 3% origination fee paid in advance amounts to 20.29%. If repayment is delayed by only three months, the average rate paid to the lender is reduced to 15.54%. This is one of the drawbacks in the concept of the point system as an advance payment. Actually, the points are a fixed advance charge against a series of events that are variable or unpredictable. See Appendix XI for table of risk and return yields.

The calculation of median interest upon which lenders of construction money base their loans is pre-determined by a theory of cash flow. There is an equal draw-down of proceeds each month of the original term, as estimated. If the time of completion of the project becomes delayed, a construction lender assumes that any additional funds are being obtained from other sources. This may be your buyer's permanent mortgage or your own arrangement for a take-out. Remember that the median interest refers only to the original amount and the original term. Any delay in payment is only on the original amount. The lender has set aside a specific amount of money for a specific amount of time to be reinvested again immediately for a similar time period. The final repayment of the construction loan influences the effect that your lender's points have on the loan. The points have increased the interest income to your lender. The rate is increased on the total amount you will be paying for the money, in accord with the number of points.

Cash flow earnings diminish if the repayment of a loan is delayed. An overall yield declines with time. A lender loses money if construction is delayed, therefore planning for this possibility with a lender is important. Both you and the lender lose money if the construction period and the loan period are not the same.

One of your alternatives when the construction lender's money becomes due is the necessity of resorting to the trade-in concept.

If marketing problems had been anticipated in your overall planning, you and your permanent lender, or in some circumstances your construction lender, may have covered such possibilities in previous discussions. The possibility of signing up for the permanent mortgage yourself, or providing for the permanent loan to be taken "subject to" or assumed in an equity exchange, was no doubt taken into consideration by the mortgage lender. Actually, an exchange for an older property can often be the source of operating capital in the absence of a substantial down payment transaction.

In an exchange you obligate yourself to a standard permanent amortized loan with provisions for a buyer or an *exchangor* to take over the existing loan. An exchangor (or trade-in) buyer of your newly built house has an approximate $20,000 equity for the down payment if you trade equities. We can call him Mr. Barter. The older home is worth around $40,000. If you take this equity on the old house and apply it to a trade on your newly constructed house, you will be "frozen" until the old house is sold in turn. In the meantime, however, you can raise substantial operating capital on the old house. The existing loan on the old house is now paid down to a balance of $20,000. On a trade-in for resale, expenses would involve prorations, sales commissions, title search costs, loan fees and pay-offs. On the other hand, if an 80% interim trade-in loan is made based on $42,500 rather than an appraisal of $40,000 on the older home or, say, a minimum of $34,000 on such refinance—then an amount of $14,000 would be unfrozen, leaving a balance of $6,000. Only $6,000 would be tied up until the old house could be disposed of.

Depending upon the condition of Mr. Barter's older home, its location, general marketability, and other factors, it may pay you to credit Mr. Barter's trade-in allowance made to around 85% or 90% of an estimate of value based on a market sales survey. In this way you would offset a sales commission on future resale.

Since builders are businessmen who operate in capital goods, you are simply making sure you are covering all bases.

A trade-in may be feasible on an eventual resale of trade-in property, even with brokerage fees plus a few dollars or so in repairs, and possibly a 2% loan fee on an interim trade-in loan. In this case, a lump sum estimate must be accounted for on the refinance and resale of the old house. Logical approaches include a dependable appraisal of the old property, another expense, and may include delays in obtaining an FHA or VA appraisal. If an independent appraiser is used on the old house, the value conclusion is a professional opinion and should not be considered the sale price. It is only an estimate on an equity exchange. But in the case of your newly built home, the value is more firmly determined. The market price for labor and material on the new construction is known — it is all on the books and has been paid in full. It is known what the building site had cost because it was purchased in order to build—a purchase backed up by a complete market sales survey. Although costs are not always value, there are no depreciation factors of a physical nature on the new structure, unless an obsolescence factor was built in or there is misplaced depreciation or other such factors. Valuation based on sales of property comparable to the old property would not be as reliable as your value estimate of the new property.

Remember in your trade-in deal that most potential sellers of their own property have over-inflated ideas of the value. They have not been made conscious that depreciation and obsolescence are extremely important and that it is land itself that is the increasing factor in realty valuation. This human element is important in the deal. The details of the trade-in depend upon the value placed on the old house regardless of whether a professional appraisal is made or not. The estimate made by the builder as to its value is just as important. A professional appraisal may not convince all parties that the deal is a good one. This is no *barter* situation. The issue is people, not the commodity. How badly does a prospective exchangor cf an old home want or need the new housing? How badly does the builder want to thaw out his frozen asset and get on to other projects? Title swapping and equity exchanging are personal with the buyers and strictly business with the builder.

The Trade-in Plan

In the face of a good substitute deal, a proposal is made in writing on the basis of an offer or counter-offer. An exchange agreement is reached based upon $20,000 equitable credit minus finance fee, commissions and other estimates of costs attributable to Mr. Barter's house.

Since the future selling price of the old house is uncertain, a net figure is used for a base price. As a builder, you are unconcerned at this point as to the selling price obtained by the listing broker (if you use his services). You must net $40,000 after the selling commission is paid in order to consider trade-in financing in advance of the future resale of the older property.

The agreed price based on a 6% commission ($40,000 divided by .94) equals $42,550, or $42,500 rounded. This is a base figure for an 80% trade-in loan and for a listing price for resale later.

(a)	LVR 80% interim trade-in loan @2% service fee	$680	
(b)	Pay-off of existing loan of $20,000 @180 day penalty (7%)	700	(round
(c)	Title policy or abstract plus escrow or attorney (estimated)	225	figures)
(d)	Repairs for upgrades to marketability, such as termite work	300	
(e)	Brokerage commission @6% (varies as to area)	2,550	
		$4,455	$42,500
			$38,045

Since expenses are 10.5% of value, as estimated in advance of the resale, an allowance of 89.5% of the estimated value is offered for the house owned by the Barters, or $18,000 credit for equitable down payment on the trade-in of the new house. This means that the Barter family will take over the new house loan to $32,000, receiving $18,000 credit at the trade-in purchase price of $50,000 for the new house.

If you the builder had been met with opposition to the old house trade-in plan, you would have to pay resale expenses on $40,000. This would call for raising the trade-in sale price to the exchangor on the new house to $54,400. The exchangor would then assume the mortgage to an adjusted balance of $34,400 still with $20,000 credit on the equity exchange.

In view of the costs involved, the builder can simply offer the trade-in purchaser $38,000, or 89.5% of the estimated sale price of the old house. This means $18,000 credit toward the sale price of the new house with the buyer's permanent loan approved at $32,000.

Banks and other lending institutions have developed real estate-secured second liens under the recorded designation, "second trust deed or second mortgage." Although they may not charge a borrower points or one-time costs in their equity funding, builders interested in trade-in loans may resort to these secondary finance procedures up to 80% of value of the real property without actually refinancing the first mortgage. A few dollars could be saved in avoiding the complete refinancing of the old house with this method. But a basic amount equalling the lump sum received by equity funds, namely the $12,000 2nd lien, would still not be adequate to unfreeze the balance still tied up—the $6,000 representing all of the working capital still remaining. Even then, the interest would be high on a bank's or other institution's secondary financing, not to mention the amortized payments on a 10 or 15 year term. There would be no flexibility in the rate of pay-off as found in private money-lending circles.

A short term, interest only, second mortgage private investor of hard money is anxious to consider lending you the builder $4,250. This represents one-half of the equity remaining, based on a private lender's estimation. You need only indicate to the private lender that, in lieu of an actual sale at this time, you may be able to obtain an 80% loan to value trade-in loan based upon the institutional lender's appraisal of $42,500 until the old house sells. Your proposal now looks like this:

Institutional appraisal if based on brokers' listing	$42,500	
Institutional trade-in interim loan @80% LVR	34,000	
Estimated equity based on new interim loan	8,500	
Additional equity funding (second loan) @½ of equity	4,250	– (private lender)
Amount still frozen on the trade-in. Final sale proceeds	1,750	

Proceeds of $12,000 from the new house loan will add to $18,250 funding of the old house. If this plan materializes and the figures are "in the ball park," working capital is now available for a new plan. Success of this plan depends upon the market, the individual builder's viewpoint and circumstances, variable lending climates and geographical activities of the housing market. The secondary mortgage financing interests also are very important. Figures 8-1 and 8-2 indicate the mechanics of leverage used in our trade-in example.

Trade-Ins and the Lender's Viewpoint

Until the enactment of the Housing Act of 1961, federally chartered savings associations had to use general loan plans to provide builders or brokers with any trade-in loans. These were nothing more than installment loans creating problems because principal amortization was required. If a straight loan were made (interest on constant principle), the loan to value ratio was reduced considerably and was classified with a statutory "percentage of assets" limitation. Savings and loan associations, in some cases, used the plan requiring interest

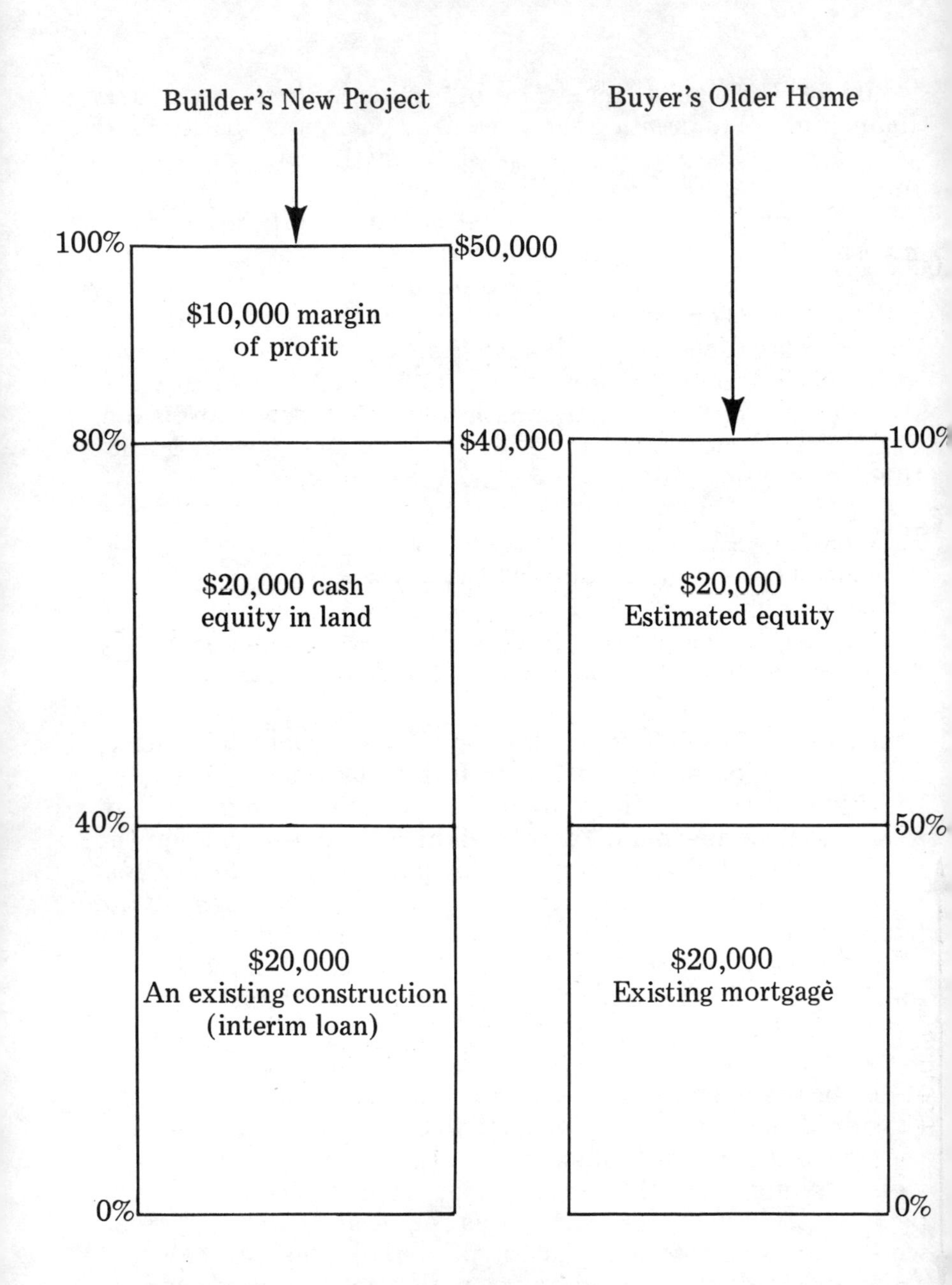

Before Trade-In
Figure 8-1

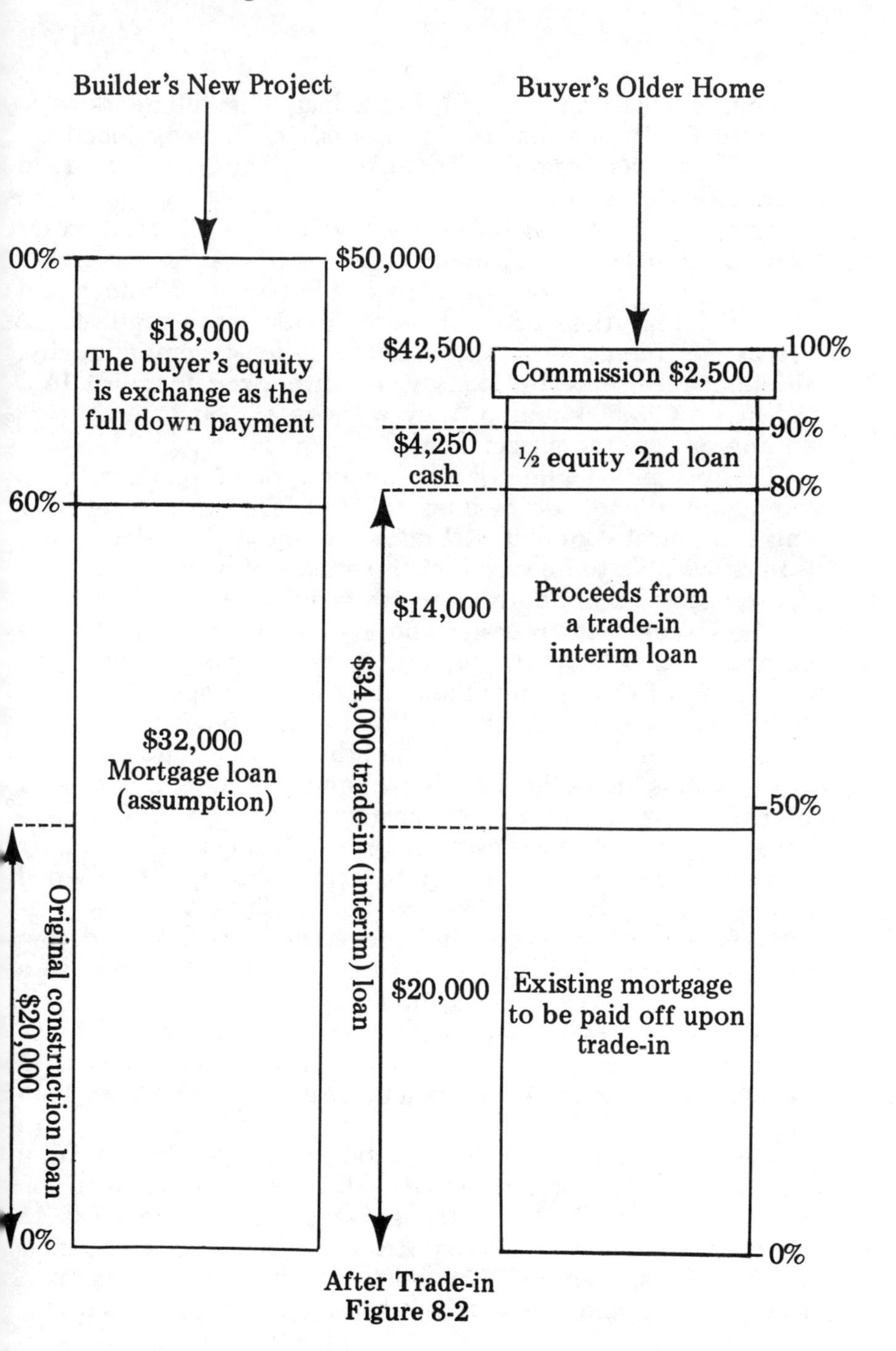
Builder's New Project
Buyer's Older Home
00%
$50,000
$18,000
The buyer's equity
is exchange as the
full down payment
60%
$32,000
Mortgage loan
(assumption)
Original construction loan
$20,000
0%
$42,500
Commission $2,500
100%
90%
$4,250
cash
½ equity 2nd loan
80%
$14,000
Proceeds from
a trade-in
interim loan
$34,000 trade-in (interim) loan
50%
$20,000
Existing mortgage
to be paid off upon
trade-in
0%
After Trade-in
Figure 8-2

payable at least semi-annually but principle installments were deferred for 12 months. During periods of housing inactivity the builder was handicapped further in the absence of an interim finance plan.

The Housing Act of 1961 contained a new authorization for federally chartered associations to invest 5% of their assets in non-amortized loans on homes or combinations of homes and business properties, with a maximum 18 month maturity, if interest payments were to be made at least semi-annually. Although these interim loans were authorized for 80% LVR and had a $35,000 ceiling, they were restricted to a specific lending area of the association.

This procedure allowed a revolving type of loan similar to floor plan loans for automobiles, taking these loans out of their long term mortgage status. Under the automobile floor plan, dealers are able to finance their inventory of new cars as well as used cars on a month to month basis with a blanket loan.

The issuance of trade-in type loans is a small part of the mortgage lenders' portfolio. But as trends occur toward an oversupply of new homes, the builder needs a special method of turnover. Since the innovation of the interim loan for trade-ins with a maximum 18 month maturity, the need for such a loan is sporadic. Variations in interest rates, as well as competition in the mortgage loan business for good mortgage loans, means that the construction market must be constantly watched. Aside from wanting to remain known as a good source of funds for the family loan market, the savings associations must assume a public image as an accommodating institution. Not much money is made on trade-in loans, so a hefty fee is sometimes charged. The interest rate is probably competitive enough, but unless there are enough of these loans with a constant turnover and with a new fee each turnover, a low interest rate is mainly an instrument for good public relations.

Institutional lenders are encouraged to explore several avenues of mortgage investment through hard competition from other lenders. One of these avenues is the trade-in. As the real estate cycles fluctuate, the need for the trade-in is affected. If the demand exceeds the supply of housing by a strong margin and the supply of mortgage funds is plentiful, no builders will need this type of mortgage. The real problem in the past has been the complicated method of unfreezing the equities of the homeowners. In a market that indicates plenty of equity in many homes, but a lack of construction activity, the homeowner cannot be blamed for hesitating to buy new property at a fair price and then end up owning two mortgaged properties. Also, possession, even when there is plenty of new housing to choose from, may be a complicating factor when there are two separate and distinct sales.

Some Mortgage Alternatives

In the 1970's several different kinds of mortgages were proposed which would augment the standard constant-payment, single interest rate amortized loan. Among them were the following:

Graduated Mortgage Payment Plan: In this plan, mortgage payments start at a lower than normal rate and increase, theoretically at a slower rate than family income. In a graduated payment plan, the amount of under-payment is added to the balance owed, thus resulting in rising interest payments. This plan should benefit young families just starting out. Example: A hypothetical $30,000, 30 year, 9% mortgage, in standard form as in the past, would cost the borrower a monthly payment of $241.00 for principle and interest over the entire 30 years. But under the most popular plan of graduated payments, the initial amount would be sharply lower—$182.00. The payment would increase 7.5% a year for 5 years, reaching $262.00.

Variable Rate: The interest rates rise and fall with changing money-market conditions. The mortgage lenders say that this arrangement can stabilize mortgage money supplies in times of rising interest rate. It does this by providing the lenders with more funds at such times, thus enabling them to pay higher interest on lending associations' savings accounts, and to attract or hold on to more deposits.

With the variable rate approach, the payment would start at $241.00 per month, the same as with the standard loan. However, assuming an increase of ½ a percentage point each year in the interest rate (the maximum annual increase allowed in current variable rate plans) the payments by the sixth year would reach $294.00.

The key difference between these two plans is in the unknown rate of increase in the variable rate plan as against the known factor of escalation in the graduated mortgage plan. Lenders claim that under the variable rate plan, loan rates could also come down.

Rollover Mortgage: Some states permit their mortgage lenders to try new ventures in addition to variable rate mortgages. One modification to the variable rate loan is called the "rollover" mortgage. This plan holds interest rates steady for five years at a time. It is automatically renewable every five years at the buyers' discretion at a rate tied to an index of the cost of real estate funds. But the monthly repayment rates are usually refigured on a longer term amortization schedule. In the event the borrower and lender do not agree on the new loan terms, the borrower has the option to find financing elsewhere and pay off his current obligations without a prepayment fee.

Reverse Mortgage: Builders and brokers may not find this type of mortgage of vital interest to a sale of their product or to a listing, but it should be included in any discussion of new innovations in the mortgage world. The reverse mortgage allows older families to take out a mortgage loan which is disbursed to them on a monthly basis, rather than in a one-time lump sum payment. An income of considerable size for an older family would no doubt be an answer to the problem of what to do with equity built up over the years. Now, and for many years in the past, the existing mortgage stands between a potent asset and a roof over the mortgagor's head. To sell means paying rent or reinvesting in another home. This reverse mortgage will serve two purposes, income and residence. The only remaining question for the mortgagor is "how long will I continue to live." The money may let him and his spouse keep their house, whereas they would otherwise have to sell because of maintenance and property taxes.

No matter what mortgage plans are formulated to accommodate the number of people who wish loans for purchase or refinance, money is still a requirement in equity on purchase of real estate. A part ownership or cash is needed for that initial qualification necessary for an institutional loan.

It is needed to cover the profit to be made on the building project. It is needed for any realty transaction in the transfer of property. If not cash, then a negotiable instrument or item of value for conversion to cash must be furnished by the borrower. Something must fill the gap between the sale price and the new or existing loan. Savings is usually the key. But it is basic salesmanship to have a varied set of selling plans other than simply cash to the loan. A smart builder can obtain his cash if he has investors standing by to pick up any purchase money, second mortgage or trust deed at a specified discount commitment in advance. Once a builder obtains offers, he can more easily promote the sale if he has more than one type of mortgage commitment for buyers with low down payments.

Families are not all the same. They all have different housing needs and they have different financing preferences. What is needed is the ability for lenders to offer some of the flexibility found in some brokerages, such as the contract of sale used for many years, as a substitute for the immediate down payment. Other examples would be the ground lease approach or purchase contract similar to the Department of Veterans Affairs plan in the state of California. In ground-leases buyer owns the structure he lives in but only leases the land. Therefore the mortgage on his home remains small and affordable until enough equity is built up for the purchase of both land and building. There exists now a ''portable'' mortgage which gives a homeowner the ability to take his present mortgage at its current rate along to his new house. Thus the borrower can transfer his mortgage to another home during a period when rates on new loans are higher.

Mortgage Transfers

In a majority of cases, mortgaged properties that are sold are refinanced. In our builder's original permanent finance plans mentioned earlier in the book, he had convinced the lender of his honest intentions and had even established a payment schedule. Now, in selling what he has built, our builder has a problem of modifying or adjusting the existing mortgage. A different balance appears, since our builder has established credit he will apply to the new owner's equity.

Since most sales of mortgaged properties are refinanced,

the existing balance is usually paid off from new proceeds of the loan. The buyer, depending upon his down payment, places a new mortgage on the property. There is hardly ever a problem for the seller, who only has to pay the amount of the debt to the lender out of the proceeds of the new funds advanced—and pay a penalty for doing so. But, in this case, how is the lender going to handle this transferred mortgage for his protection?

Most problems appear when a mortgaged property is sold without provision for immediate retirement of the old mortgage. This occurs when a property is sold without mention of the existing encumbrance; it is sometimes sold ''subject to'' the mortgage. Most times it is sold with the deed expressly providing that the purchaser ''assumes and agrees to pay'' the mortgage. This is the type of sale transaction that will be discussed here, in respect to the mortgage creditor. It is a fact in mortgage law that the mortgagors' interest in encumbered realty is transferable without question. Since there are many kinds of separate transfers of realty, we will limit the discussion to market sales transactions between buyer and seller in the open real estate market and to the mortgage continuity problems involved.

A mortgagor has an absolute right to sell his interests in real estate and a mortagee cannot prevent this right. But there are certain options which can be exercised by the mortgagee, such as acceleration clauses. However, such clauses are not considered of great value except in cases of default in payments. To accelerate a maturity simply because of a sale to which the lender did not consent is defeating the purpose of mortgage lenders and therefore is seldom exercised. The customary mortgage made by a savings association involves two elements: (a) the mortgage lien upon the property, and (b) the borrower's personal obligation to the lender for the debt. In the event of default, the mortgagee may look both to the property and to the personal debt for satisfaction. If, upon foreclosure, the value of the property proves inadequate to meet the debt, the deficiency is claimed from the borrower upon his personal obligation.

The buyer of the newly built property cannot see any reason why he should not be allowed to merely pay down the existing mortgage and take the balance of the mortgage ''subject to'' the mortgage amount remaining on the property.

No obligation or legal responsibility is implied by his buying property ''subject to the mortgage,'' other than that the mortgaged property must be the primary security for the debt. In order to indicate personal liability, it is imperative that there is a promise in definite terms to pay the mortgage. Taking over the existing mortgage is an economic move, as far as the buyer is concerned. A new application entails a complete pay-off, new title searches, reconveyances and possibly different terms, since there is a different mortgagor involved. The buyer cannot be blamed for avoiding these expenses.

As a general rule, the buyer of mortgaged property assumes no personal obligation whatsoever, simply by reason of a ''subject to'' clause. Only if the buyer negotiates an *agreement* with the seller to pay the debt upon foreclosure is he personally liable. It is for this reason that the device covering an ''assumption of mortgage'' is used in a case of a transfer, whether or not a balance has been thereby reduced in the transaction of sale involved— the buyer must agree to pay, or assume the mortgage or mortgage debt. The purchaser then becomes personally liable for the amount of the debt, a liability which can be enforced by the mortgagor (seller) with whom the arrangement is made.

While the agreement to assume the mortgage may be expressed or implied, written or oral or separate from a deed, the lender in our example requested a separate three party agreement. This is because the assumed agreements were not incorporated into the conveyance. The assumption agreement, as shown in Figure 8-3, illustrates the understanding between the parties. The mortgagee consents to the sale and a substitution and agrees to accept payments from the buyer. It specifically states that the agreement releases the *original mortgagor* from liability.

Subsequently, if the lender is thoroughly convinced that his security is protected and the buyer is unquestionably reliable, the original mortgagor can be released by a document. This separate and distinct document is shown in Figure 8-4. It releases the original mortgagor and replaces him with the buyer, indicating the buyer's full liability.

FHA FORM NO. 2210
Rev. 9/67

Form Approved
Budget Bureau No. 63-R0804

U. S. DEPARTMENT OF HOUSING AND URBAN DEVELOPMENT
FEDERAL HOUSING ADMINISTRATION

REQUEST FOR CREDIT APPROVAL OF SUBSTITUTE MORTGAGOR

FHA Case Number

INSTRUCTIONS: This form is for use in cases involving the release of a Mortgagor from liability for a deficiency occurring as a result of foreclosure. Submit original only to FHA.

SECTION OF THE NATIONAL HOUSING ACT

☐ 203 ☐ ____________

MORTGAGEE *(Name, Address & Zip Code)*

Property Address *(Street, City & State)*

SELLER *(Name, Address & Zip Code)*

PURCHASER *(Name, Address & Zip Code)*

A. MORTGAGEE'S REQUEST FOR SUBSTITUTION:

It is requested that the above named purchaser be accepted as Mortgagor and the Seller released from financial responsibility fo a deficiency occurring as a result of foreclosure.

An FHA Form 2900, Mortgagor's Application for Credit Approval, with required exhibits is submitted herewith and the statement contained therein are true and complete to the best knowledge and belief of the undersigned.

Title Of The Above Property: ☐ Has been Transferred ☐ Will be Transferred	Monthly Mortgage Payment *(Total Principal, Interest, M.I.P., Ins., Taxes, and any ground Rent or Special Assessments)* ⟶ $	Remaining Term of Mortgage ______ Months	Face Amount of Original Mortgage $
MORTGAGE - ☐ IS ☐ IS NOT CURRENT	Date of First Payment *(Original)*	Purchaser is or will be Owner-Occupant ☐ Yes ☐ No	☐ Insured under Escrow Commitment Procedu

Date ____________________ By ____________________
Name and Title of Officer

B. CONSENT BY THE FEDERAL HOUSING COMMISSIONER

The above named Purchaser is acceptable as a Mortgagor and, subject to compliance with the following conditions, if any, and the issuance of Form 2210-1 to the Seller, consent is given to the release of the Seller from Financial liability for a deficiency occurring as a result of foreclosure in connection with the above numbered loan. Form 2210-1 shall not be executed by the Mortgagee until the sale to the above named Purchaser is concluded and conditions specified below are met.

SPECIFIC CONDITIONS

☐ The principal balance of the mortgage be reduced to $____________ or less.

☐ The purchaser deposit with you in escrow, trust or special account $____________ under an agreement whereby; (1) the funds will not be disbursed until the property is sold to an owner-occupant purchaser acceptable to FHA; and (2) if the propert is not sold within 18 months from the date of transfer to the above named purchaser, the funds shall be applied as a mandatory prepayment to the mortgage principal.

ASSISTANT SECRETARY-COMMISSIONER

Date ____________________ By ____________________
Authorized Agent

NOTE TO MORTGAGEE

A copy of this form has not been retained by FHA in its files. Within 30 days of change, you are required to submit to FHA, Form 2080, Mortgage Record Change, to the Assistant Commissioner-Comptroller, Federal Housing Administration, Department of Housing and Urban Development, Att: Receipts and Deposits, Washington, D. C. 20412.

☆ U. S. GOVERNMENT PRINTING OFFICE 1975 689-416/179

FHA FORM NO. 2210 Rev. 9/6

Assumption agreement
Figure 8-3

ADMINISTRATION OF
INSURED HOME MORTGAGES

FHA FORM NO. 2210-1

U. S. DEPARTMENT OF HOUSING AND URBAN DEVELOPMENT
FEDERAL HOUSING ADMINISTRATION

APPROVAL OF PURCHASER
AND
RELEASE OF SELLER

FHA Case Number

SECTION OF THE NATIONAL HOUSING ACT

☐ 203 ☐ ____________

Mortgagee *(Name, Address & Zip Code)*

Property Address *(Street, City & State)*

SELLER *(Name, Address & Zip Code)*

PURCHASER *(Name, Address & Zip Code)*

This will acknowledge that the above-named seller has sold the property described above to the purchaser named.

The credit of the purchaser has been examined and approved by FHA. The seller is hereby released from any financial obligation arising in connection with the security instruments executed in the above numbered case. No deficiency judgment will be taken against the seller if the FHA-insured mortgage covering the subject property is foreclosed.

If the seller should apply for an FHA-insured loan on another property, this release should be delivered to the Mortgage Lender through whom the application for such loan is made.

____
Mortgagee

BY: ____________________________

Date ____________________________

NOTE: This is an Important Document that the Seller should Retain.

Figure 8-4

ADMINISTRATION OF
INSURED HOME MORTGAGES

SAMPLE RECASTING AGREEMENT

FHA Case # ______________

This Agreement, made this day of , 19 , between , , hereinafter referred to as Lender, and , hereinafter referred to as the Borrowers, and , as Trustee (if applicable);

WITNESSETH:

Whereas the Borrowers are now indebted to the Lender in the sum of Dollars ($) (hereinafter called "new principal amount"), consisting of Dollars ($) unpaid principal, and Dollars ($) unpaid installments of ground rents, hazard insurance premiums, taxes, assessments, and mortgage insurance premiums, the payment of which is secured by a note and security instrument owned and held by the Lender, dated , 19 , and recorded in the office for the recording of deeds in County and State of , in book of mortgages, page , and

Whereas the parties mutually desire to modify the terms of payment of said indebtedness by changing the amount of monthly payments required on said note and security instrument;

NOW, THEREFORE, in consideration of the covenants hereinafter contained, it is mutually agreed as follows:

The Borrowers agree to pay the "new principal amount" with interest at the rate specified in said note on the unpaid balance in monthly installments of Dollars ($) commencing on the first day of 19 , and on the first day of each month thereafter until the "new principal amount" and interest thereon are fully paid, except that the final payment of the "new principal amount" and interest, if not sooner paid, shall be due and payable on the first day of , 19 .*

It is mutually agreed that said security instrument shall continue a first lien upon the premises and that neither the obligation evidencing the aforesaid indebtedness nor the security instrument securing the same shall in any way be prejudiced by this agreement, but said obligation and security instrument and all the covenants and agreements thereof and the rights of the parties thereunder shall remain in full force and effect except as herein expressly modified.

IN WITNESS WHEREOF, the parties have signed, sealed, and delivered this agreement on the date above written.

______________________________(SEAL) ______________________________(SEAL)
Lender Borrower

______________________________ ______________________________
By Borrower

DEEDS OF TRUST

(If the security instrument is a deed of trust and it is necessary that the Trustee execute recasting agreements, the following acknowledgment shall be signed by the Trustee.)

THE TRUSTEE has executed this instrument to acknowledge his (its) assent thereto and agrees to continue to act in such capacity under the terms as modified herein.

TRUSTEE:

____

____

*This date cannot exceed by more than 10 years the maturity date of the original note.

(Add acknowledgment, if required)

2/77

Recasting agreement
Figure 8-5

In an assumption agreement to a mortgage debt the laws of the state in which the assumption contract is performed control the procedures for all parties involved. HUD/FHA forms are sample guides — only for information purposes. Conventional-type loan forms can be similar, however.

Since mortgage loans are personal from a lender's standpoint, a new mortgagor involves a new loan balance owed to him by new people. Continuity of a mortgage is not a simple matter if the ownership changes. In our example, a buyer agrees to an exchange of equities (with adjustments). But he finds that all requirements of the mortgagee (lender) include not only a three party agreement to an assumption, but a "modification agreement." Also, this serves notice that all parties are in possession of specifics as to the balance of the loan due the mortgage creditor. This HUD/FHA form appearing in Figure 8-5 serves as a guide to conventional loans as well. Assumption agreements are optional in many cases, depending upon the requirements of the individual lending institution and local laws. An attorney should provide counsel in this type of agreement.

Builder's Trade-In Outline

If our builder had *not* followed through with a complete new mortgage the $20,000 would remain in his hands as his original debt. The marketing would still be undone, however. The committed $40,000 permanent type loan was approved, processed and actually recorded. An assumption document in an amount of $32,000 was required for the buyer of the new home. The builder's construction loan debt was required to be paid off through the Barters' purchase. The assumption agreement as well as the loan modification agreement show that the balance now owed is the $32,000 on a permanent mortgage loan.

The builder's final lump sum cash after the trade-in but before the sale of the older home looks like this:

Proceeds from $32,000 commitment assumed by Barters	-	$12,000
Proceeds from $34,000 trade-in (interim) loan on old house	-	$14,000
Proceeds from $ 4,250 hard money second mortgage or trust deed	-	$ 4,250

As indicated earlier, if the older house eventually sells for $42,500 there still remains the $1,750 as final proceeds to the builder.

Here is an outline of the builder's financing venture, beginning with the original refinance of his home to provide building funds—primarily for a site.

1. Soon after an application was initiated with the savings and loan association for refinance, the builder submits an offer to the landowner for $20,000. This contingent deal is accepted by the owner of the building site. See Figure 6-1.

2. A bank located by the builder makes its commitment for a construction loan on this lot. The bank's current policy prevents it from furnishing a take out or permanent type loan, leaving the builder to find his own permanent loan. The interim loan is out for one year, with interest payable at 11% interest only monthly payments of $183.33. The loan fee is 5 points (5% of the $20,000 applied for) or $1,000. Related fees include an appraisal, $60.00; inspection fee, $110.00, or $27.50 for each of four inspections; escrow, $85.00; drawing documents and accounting services $30.00; fire insurance, $150.00 during construction. In addition there are the usual recording and title search charges, plus proration of the taxes.

3. In the same month, another savings association contacted by the builder offers an interim-permanent type mortgage on the building lot which includes not only the construction costs but an option to increase the mortgage up to 80% LVR in the builder's name upon completion. (This is up to the loan committee upon re-appraisal of the finished product.) The interest rate is 10% at 3% origination fee for 6 months. But this offer is more attractive than the previous one because (a) there are to be no installment payments for six months, and (b) the mortgage is to be one encumbrance on construction and take out. This lending association was therefore chosen for a construction-permanent (take out) loan and funds were disbursed for the project. The builder's $20,000 cash paid for the land was his existing equity.

4. The building project is completed.

5. Although not buyers with cash for a down payment, Mr. and Mrs. Barter negotiate with both the builder and the lender of the new house and take over the builder's loan.

6. The lending association used for the builder's financing is now a part of the trade-in, but here the builder negotiates with this lender for the Barter's old house so that his working capital (and profit) can be obtained. (See trade-in charts, Figures 8-1 and 8-2.)

7. With the mortgage on the new house fully assumed (see Figure 8-4), the lender processes the interim trade-in loan on the home that the builder takes in trade.

Loan Assumption and Buy Downs

Under the U.S. Supreme Court's De la Cuesta ruling, loans originated by federally chartered savings and loan associations are not assumable by the new buyer. The rule for state chartered S & L's is more complex. Loans made or extended before August 1978 or after October 1982 are probably not assumable. But the August 1978 date is subject to interpretation by the courts of your state. Note, however, that even mortgages originated between August 1978 and October 1982 may be assumable only if the assumption is completed before *October 1985.* The fact that a state chartered S & L switched to federal charter would have no effect on the assumption, even if a due-on-sale clause is written into the loan contract. If assumption is not possible, consider negotiating with the lender for a *buy down* on the loan.

Here's an example of how a buy down works. Assume that a $40,000 second loan is needed. A lender wants a 12% loan amortized over 30 years — $411.45 monthly. Your buyer insists on a 10% loan at $300.14 monthly and will pay full price for the home if that loan can be arranged. You, the builder, buy down the loan by paying the lender $5,800, leaving the effective loan balance $34,200 ($40,000 minus $5,800). The lender then makes the $40,000 loan to the buyer at 10%. But the lender makes 12% on the loan because of the seller's buy down. Your $5,800 payment plus the buyer's payments leaves only a $111.31 difference. To buy that difference down for 3 years costs $4,007.16, or $111.31 x 36 months. If that buy down can sell a $400,000 home, it's well worth the cost.

9

Tract Financing

As a builder, you have acquired ten acres of raw land at $8,600.00 per acre, based on equity funding of your existing property. Leverage is part of that reinvestment. A period of two years is all that you can devote to that raw land investment as land value increases toward the day of resale. This is plenty of time for a subdivision feasibility analysis. One of your first questions is, "Who makes subdivision tract loans?"

Mortgage bankers, savings and loan associations, insurance companies with their loan correspondents, syndicates and partnerships all are geared for such an investment. Sometimes the landowners themselves remain as participants. With this surplus of loan sources, you should look carefully to find the best subdivision financing for your project. You will have prepared yourself to pass all qualifications, such as engineering plans and marketing considerations. Your first practical step is to create the tentative tract map as the main tool in selling your idea to the proper lending organization. As you make your contacts, it should be kept in mind that the thrift institutions are the largest type of real property financing institutions in the housing field and in their own communities. Unless your development project is remote from an urban area, some savings association will probably originate the loan and elect to sell it from their own portfolio of investments. They do not wish to become mortgage bankers, as such, but merely want to meet the financing needs of the community. Meeting these needs is actually their responsibility, as specified in the federal and state laws under which they are chartered. At one time a fifty-mile lending radius was the rule with most savings groups, but the Federal Home Loan Bank Board permitted more flexibility for convenient inspections and in the personnel requirements for subdivisions. But aside from savings and loans, the most generous LVR at the lowest APR will always be the ideal loan regardless of its source. Preparation is the key in the approach to any lender. Next to your good credit, a good tract map is the most valuable possession of all. Even the financing itself is only one of the tools, important as it may be, in putting together a subdivision tract.

The Developed Lot Loan

The "land layout" should have the atmosphere of the countryside. Your prospects will probably be people who have come out to the suburbs to get out of the city. Trees should be left standing and roads should be laid out as winding through the area but not so much as to create potential problems for traffic. The sharp, hilly mounds should be removed and all gullies filled in, but try to arrange a gently rolling terrain for attractiveness. Lot sizes should be appealing but the best lot size depends upon the type of development. Lot size of thirty or forty feet wide and one hundred feet in depth is typical of the inexpensive development. Better developments will have plots as large as three to five acres.

Savings and loan associations favor developments encompassing single-family home lots or sites. Multiple unit plots, or plots for two to four family dwellings, may require negotiations with lenders other than savings associations. Multiple unit developments should be located on the fringe of the more populated areas. Mortgage bankers with insurance money are the "packagers" for this kind of development. Large syndicators or partnerships also favor multiples.

If the development is near an urban area and you are only acting as a subdivider, then you may be simply selling the lots or promoting the project as a package to other developers or builders. This is usually the procedure when all cash is paid for the land, the tract map is filed, the improvements for streets are in and all installations of public facilities are made by the municipality. The final inspection is made by the local government.

If a lender is involved up to this point, further requirements may be mandatory to bring the tract up to full lot improvement status. As a matter of procedure before requesting a commitment from any lending source, your full subdivision preliminaries should be completed, such as the readiness of the final subdivision map for recording. In this respect, the following steps are usually necessary in many counties and to a lesser extent in surburban areas. Possibly your reinvestment of raw land may be subject to these preliminaries in a subdivision process.

The Subdivision Process

- The subdivider or his engineer discusses the proposal with a staff member of the appropriate governmental planning group. This is prior to preparing a tentative map. Later the staff member studies it for compliance with basics.

- Upon completion of a tentative map, the subdivider submits 20 or more copies to the planning group. The owner's statement along with the filing fee will be submitted for processing at this time.

- There may be a matter of weeks between filing your tentative map and its consideration by the subdivision committee. During this time the maps will be distributed to the committee members, school districts, adjacent cities, the highways department, the state real estate commissioner and possibly other agencies as well. In some localities, this process could mean fewer copies and a shorter period of time. Field inspections and other investigations are made during the time before the members submit their final report.

- A date is established for the subdivider and his engineer to meet with the subdivision committee which may include representatives from departments such as the highway, flood control district, storm drain, sanitation, mapping, and the building and safety division of the county engineer. The meeting could include members of of the local health department, parks and recreation department, forestry or fire warden's office.

- The subdivision committee's recommendations are presented to the regional planning commission for final approval.

- The subdivider receives his copy of the conditions of approval and, in turn, instructs his engineer to prepare a final map of land improvement plans.

- The final map is distributed once again and checked. Each county department reports to the county engineer regarding its compliance with the conditions of approval.

- The board of supervisors or the appropriate governing body in your area approves the final map for recording after agreements for further improvements are made. The board clerk will certify the action.

The subdivider's tract is ready for the sale of the lots. Either the subdivider's own promotional facilities are used, or the package can be sold directly to a builder and construction developer (see Figure 9-1). This package includes an unimproved raw land subdivision tract map as it was recorded. The crucial point in the development comes now, when proper financing is needed. Preliminary shopping for funds should have been well farmed out or negotiations with interested lenders should have already reached the stage of a preliminary commitment. You may face a tight money situation at this late date, or your own ownership problems. In any case, your tentative tract map is part of your feasibility analysis whether you yourself or others continue the development. Since no lender will commit funds without complete plans of the

FOR SALE
RECORDED MAP
Large single residence
24 lots — Chatsworth
San Fernando Valley
Owner (213) 000-0000

This owner expects to participate in the development as a subordinated party by finding and selling to a builder. The acreage has been mapped out and covenants prepared for insertion into deeds. He probably has not made a time-line appraisal at this point, hoping to feel out the market through builders.

Figure 9-1

tentative tract map, the final recording of a tract is meaningful to lenders. They consider your intentions to be serious only *after* recording, not before.

Negotiating The Loan

If your acreage is in a suburban area and you can pay the entire cost of the land in advance, as well as the expenses of getting the property in shape to sell, then obviously no leverage problems exist. On the other hand, you may have purchased the land by giving back a substantial mortgage and paying only part of the cash. In the case of a mortgage taken back by a landowner, this introduces some risk into loan negotiations, since you pay interest on the credit besides paying for the costs of the improvements. As you sell the lots, a specified amount of the sale price of each lot is paid over to the mortgagee (landowner). As each lot is sold that portion of the *blanket mortgage* is released.

You may have entered the picture as a full developer or builder as a partner, corporate member or syndicator at the time when the tract land is to be developed. In this case, you can obtain an outside construction loan. The original landowner must have already been bought out completely, or any interests still retained by a former owner can be subordinated to the new construction loan. These retained interests are undesirable, and are usually eliminated by paying off the former owner either through the new first mortgage for construction or by other means. And few lenders will finance your building plans under such conditions—interests retained by owners makes for a riskier loan on the land.

Marketability is important to lenders. The groundwork should be laid for studies typically used in a subdivision plat. You are in the wholesale business when you create a subdivision for future dwellings, or layouts and development for resale of any type of real estate. The ultimate goal is to create a commodity to be sold as a retail item. Your financial backers must look at several distinct and important details. The degree of risk that your proposal offers depends on future improvements and not on the present improvements to your tract. The filing of your tentative tract map with the governing authorities has only met the initial requirement. Now a lender will look at the subdivision layout and its proximity to neighborhood facilities and proximity to an urban area.

The improvements you have made at this beginning stage should answer these questions and many more: What are the provisions for water? For sewers and septic tanks? Have you planned for streets, curbs and lighting? Which of these are to be put in? In some cases there are legitimate reasons for waiting for street installations. You may not want to install them until all heavy equipment use is completed. In some cases the lender may hold back a portion of your loan to cover some improvements. Ideally in the eyes of the lender, all improvements to the tract will be in and paid for. A loan to build homes could face a special assessment, in which a lender is obliged to advance further funds on the individual loan. If you do not anticipate this in advance, then the purchasers of these homes may have to bear additional expenses in order to provide themselves with conveniences they should have taken into consideration in buying the property. This is not a necessary requirement for your loan, as some conveniences are not possible in some areas. But all property is worth more with modern improvements. And since projects with good facilities are less risky ventures, you may be able to obtain lower loan rates when improvements are installed before the loan is taken out.

Subdividers fall into two general categories regardless of the size of their projects: 1) the builder or developer whose acreage is at the border line of the urban community, and 2) the one who acquires acreage in the suburban area. The first type of builder has a more complicated situation. Acreage close to the center of a population or at the edge of an established community must apply the land-use concept that prevails in the area. On the other hand, the developer in remote acreage has much more control over his project. He may install roads that wind along the natural contours of the land. He can divide his lots into any size he wishes. Your lender will be just as enthusiastic in a remote project as in an urban or suburban one, providing you stay within acceptable limits of common sense. If the loan is to be based on a "quick profit" development, the risk is high. Few lenders will participate unless all the marketing angles are thoroughly worked out. Lenders think in terms of *commuter time* rather than miles to the project. The remote subdivision can be profitable and a low risk if the developer takes advantage of potential buyers' desires for nearby recreation.

The project near the borderline of the urban community is susceptible to special hazards not found in the country and is usually the last development to take place. But the scarcity of land near the city, traffic hazards, commercial or military air traffic, freeways or busy main streets, noise and fumes often do not adversely affect the building trade. Cities continue to draw enterprising developers. In periods of adequate mortgage funding, some institutions actually prefer this type of urban security for their portfolios.

Sometimes lenders are simply reluctant to invest in a project for apparently no good reason. In this case, the question to ask yourself is, Does the project have potential, or am I looking for some backer to take a chance? Find out why your lender is reluctant. He may be telling you something he doesn't really believe himself. If he indicates to you what is wrong with your proposal, use tactful persuasion and make him prove his point. If he remains reluctant and you are sure of the project's potential, make an appointment with a higher authority in the institution. It might be one of the best investments the lender ever made, but he may not want to lend the money because of a recent bad loan made in the same area. In any case, find out the truth even if it hurts. You will then know how to handle the next lender when you present him with the same deal. Always transmit confidence in your project even though you may be new in the field. A talk with the appraiser can sometimes make the difference, too.

Lot size is important in the planned area. Try to get as many lots as possible from the acreage. However, in a developer's eagerness he may sometimes reduce the size of a few lots in order to obtain one more. An appraiser may be quick to reduce the value of these smaller lots. There is never a standard rule for lot size but it must be appropriate to the development. Wooded sections should be preserved. If necessary spend more to retain trees, as they can pay for themselves as a marketing angle.

In tract financing, your lender will probably be very sensitive to the protective covenants or restrictions to be placed in the individual deeds to each property. He may take steps to make certain covenants a part of the loan contract. This guarantees the quality of the subdivision's future. Here are a few of these covenants.

1. Minimum coverage of the lot and setbacks.

2. Minimum improvement size and height.

3. Community recreational requirements.

4. Restrictions concerning commercial developments, pets and animals.

5. Fences.

The above restrictions, or covenants, may be regulated by local governing authorities, in which case the lending institution will check your awareness of zoning regulations which affect these covenants. Pride of ownership will play a part in the future growth and value of your planned subdivision.

After the completion of a subdivision, you may plan to extend your operations to include the erection of a recreation community or shopping center if zoning allows. Invariably, you must first gain momentum and enthusiasm from actual sales. Lots improved with attractive homes have more to offer than vacant lots, and are easier to sell. Buyers will purchase more quickly if some settlement is already established in the subdivision. In some cases, subdividers may wholesale the lots to outside operators, builders or building contractors who will agree to construct a certain number of houses in keeping with the style, price range and other limiting factors desired by the subdivider. In this approach you avoid the extra risks, work, supervision and financing arrangements involved in building complete subdivisions. You then may elect to become the sales agent, collecting commissions as you would if you sold older properties.

Financing is just one of the tools in putting together a subdivision tract. Financing, combined with legal, engineering and organizational functions, can make a project successful. Sometimes landowners handle the actual subdivision operation, the construction financing, and the carrying of the properties until they are sold. Large operators find it convenient to set up a number of subsidiary companies in order to maintain specialization and to keep finances separate. There could be one company holding title to the subdivision and later the unsold portions, another to conduct the building operations along with subsidiary supply houses. A third company would act as a financing agency, and a final company would conduct the actual sales of the homes or units. The latter could be the original real estate company itself. Whether yours is a small development—five parcels is considered a subdivision— or an extremely large one, a master financing plan must be broken down into three essential parts.

1. Physical layout of tract in engineered detail.

2. Initial financing and continuing financing to sale of last lot.

3. Advertising and sales promotion.

Time-Line Appraisals

Various regulations may be encountered by the subdivider depending upon the state in which he is operating. For you to understand the relationship your proposed subdivision may have to a loan application, here is the appraisal procedure for subdivision projects. The charters of various savings groups make regulations complex. Subdivision lending policies vary as to LVR. Appraising unimproved land, and the cost of improvements to be made on it as well as the increased value to be developed, is a matter of specialized appraising knowledge. Your institutional lender may use his staff appraiser for such work; but if the staff has been limited to appraising single building sites or building sites with improvements in place, he will probably hire an independent fee appraiser for such work. It is essential for cost estimating that the appraiser have use of data prepared by engineers who have made surveys, soil tests and the like. As it appraises raw land for subdivision, the lending institution will plan a development and sales program with you in order to estimate the flow of money during the project. In this way the lender can produce a realistic opinion of raw land value. In the eyes of the lender, time is the most important factor in land development programs in which the savers' money is at stake. The program schedule, the outstanding loan, the original terms of repayment and the credit standing of the subdivider are most important.

In the appraisal of raw land, time is vital for development purposes because money is tied up during the development while interest is being carried. If construction or sales encounter delays, you as the borrower-developer will be immediately affected. You will be paying added interest on your financing, which comes out of anticipated profits. The money and its profit to the lender must be respected, because there are other borrowers waiting in line for this same money who can yield added profit to the lender. If points are part of your loan, a time delay is actually disastrous to the lender. Points are part of the overall yield.

Land evaluation from the standpoint of income is similar to

appraising any retail item, because the appraiser must check market sales. He must compare the land being appraised to the land sold. He looks for a market demand for other subdivided land with the same "highest and best use" of raw land.

Other estimates which must be made include:

A. A cost estimate of bringing in off-tract utilities.

B. A cost estimate of improvement costs on the tract before the final sale, based on the maximum number of lots, streets, curbs, trees, and other improvements.

C. An opinion of value for all sites as fully improved.

D. An estimate of all taxes to be paid during the full period.

E. An estimate of time delays in the development.

F. An estimate of final income from sales.

An opinion of value for the raw land cannot be calculated until the above six estimates are made. For instance, in your market analysis of identical raw land, yours and the comparables are about equal, but your tract is situated ½ mile farther from the sewer lines. Therefore, your lots are depreciated in value from those lot values estimated for the comparables because of the cost of extending the sewer line.

Your company and the lending institution will develop a *time line*. This is the project's schedule of development and sales. An appraiser will rely on a work sheet upon which he can plot the proposed subdivision's development expenses and sales income for each year or for each scheduling period in which certain aspects of development and sales will take place. The time-line worksheet will provide information on when and how much money is to be spent on the project and what the interest will total over the marketing period of the property. The time line worksheet will also indicate the amount of

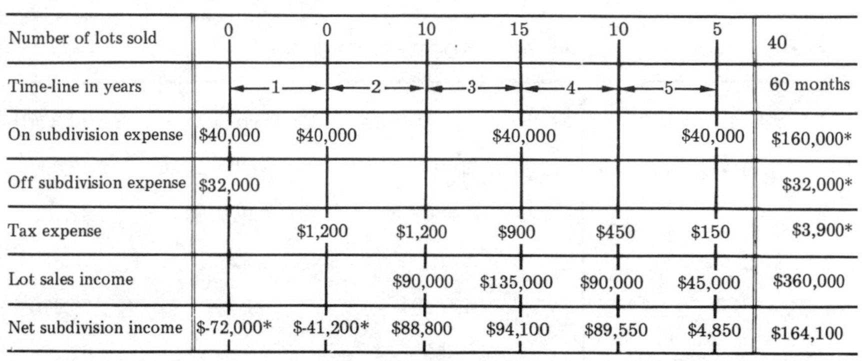

Number of lots sold	0	0	10	15	10	5	40
Time-line in years		←1→	←2→	←3→	←4→	←5→	60 months
On subdivision expense	$40,000	$40,000		$40,000		$40,000	$160,000*
Off subdivision expense	$32,000						$32,000*
Tax expense		$1,200	$1,200	$900	$450	$150	$3,900*
Lot sales income			$90,000	$135,000	$90,000	$45,000	$360,000
Net subdivision income	$-72,000*	$-41,200*	$88,800	$94,100	$89,550	$4,850	$164,100

*Expense estimates

Time-Line Chart
Figure 9-2

income that can be expected from sales, and when it can be expected. This worksheet indicates the cash input as well as the return flow of money during the life of the project.

An opinion of raw land value cannot be determined without your time-line schedule, because interest or carrying charges cannot be accurately recorded. If interest is underestimated, the raw land value would automatically be overestimated. As was seen in earlier chapters, loan appraisals done for the purpose of determining a loan to value ratio are nearly always based on a single appraiser's opinion after he obtains the basic data. Appraising to find the value of raw land for potential development is even more inexact, since inflationary cost factors are large and uncertain even with detailed cost engineering data. An LVR of 60% would be a dependable risk rate. This is the lender's capital investment for packaging, or committing, a loan for on-site and off-site improvements—before development of the acreage begins. However, remember that the value of the raw land cannot be established until the entire project, utilities, excavations, landscaping, sales and all contributions to the completed retail product are estimated. In tract financing, formulas for determining LVR as the development progresses vary from lender to lender. As an example, we can determine the LVR on a hypothetical ten acres of raw land. See Figures 9-2, 9-3, and 9-4.

PRESENT WORTH ESTIMATE

End of year	Net income projections		Discount factor		Present worth
Date of Value-O-Starting date	$-72,000	x	1.000	=	$-72,000
1	-41,200	x	0.909	=	-37,450
2	88,800	x	0.826	=	73,350
3	94,100	x	0.751	=	70,670
4	89,550	x	0.683	=	61,163
5	4,850	x	0.621	=	3,012
	Present worth of land and profit			=	98,745
	Developers profit (15% of raw land)			=	-12,880
	Estimated value raw land			=	85,865

Figure 9-3

1. Market studies estimate that the highest and best use of the ten acres is residential. Forty lots can be realized from the acreage, if 4 per acre is used in the time-line.

2. Fully improved sites reflect a total value estimate of $360,000 with the average lot value of $9,000. On the first day of each year the figures for income and expense are plotted. This simplifies interest estimates or carrying charges, although in actual practice the loan money flows uniformly throughout every year your project is in progress.

3. The period estimated for complete development and sales program is five years maximum with on-tract improvement costs estimated at $160,000.

4. Off-tract sanitary sewer line is estimated by engineer's projection at $32,000, a lump sum advance payment.

5. Tax estimates per lot per year are $30.

Date of value estimate____________ Development period__________

LAND VALUATION SUMMARY (40 LOTS)

Effective income from sales		$360,000
Development expenses:		
Improvements "on subdivision"	$160,000	
Improvements "off subdivision"	32,000	
Taxes	3,900	
Less total of development expenses		-195,900
Raw land + carrying charges + profit		$164,100
Less carrying charges		- 65,355
Profit + raw land		$ 98,745
Less profit (estimated as % of raw land)		- 12,880
Raw land		$ 85,865
Estimated value of raw land		$ 86,000

Figure 9-4

6. On-tract improvements are estimated at $160,000 in accordance with the engineers' calculations. Engineers have determined that installations can be made without disturbing the sales presentation of the program.

7. A promotional sales program is extended to a theoretical five year period, providing ample delay factor. Under this time-line plotting, no sales will be anticipated until the end of the second year when the first ten fully developed lots are sold. At the end of the third year 15 lots are predicted. Another 10 lots will be marketed at the end of the fourth year and finally the last 5 fully improved lots at end of the fifth year.

8. A bare minimum profit to the subdivider is figured against the purchase of raw land for development. This figure should expand in spite of a low time-line calculation of today's raw land value — land development estimates for loan purposes must be conservative and realistic at the same time.

9. Mortgage money from investors in raw land demands at least a 10% return per year for subdivision development for retail sale. Carrying charges, interest, points and the like are anticipated in the appraisal summary.

Raw land value can be estimated and a final opinion made if dependable facts are available for your proposed subdivision. The data put down in a format such as found in our time-line illustrations should serve as a guide for any similar proposal. The data should be then assembled and broken down into annual divisions. As indicated in the time-line chart, the scheduled expenses and income are based on an extended sales period. As expenses are paid, this money plus the raw land value is considered capital investment. The return on this capital investment is realized from the improved lot sales. As returns are realized, part of the sale price is the subdivider's profit and part is the investor's interest on the loan for the period of time his money was used. The amount of profit expected by the raw land subdivider, and the yield expected by the investor, may vary substantially depending upon the locale of the project. Profit and yield may also depend on size and quality of the raw acreage and its capacity to absorb the demand of the population.

Rate of interest is hypothetical in our example, but a 15% return for profit or 10% annual return on investors' money is not out of line for subdivision financing. The procedure for obtaining the present worth of a raw land proposal is the same anywhere. A subdivision net income of $164,100 shown in the time-line chart is not really the final figure. In this chart (Figure 9-2) a gross sales figure of $360,000 is indicated, but the raw land value and the subdivider's profit cannot be satisfied with $164,100 over a five year period. Carrying charges (interest) must be paid for the use of the investors' funds. Figure 9-3 shows the present worth calculations with these interest charges. These charges greatly affect the subdivider's profit and the raw land value. Net subdivision income totals are taken from the time-line chart and applied to the present worth discount factors. Since hard money of 10% is assumed in the loan, these factors are taken from the Single Payment Present Worth of One Dollar Tables (Appendix XIII). Each year's contribution to the present value of the raw land and the developer's profit can be determined by multiplying the net annual income and expense totals by the Present Worth Discount Factors.

This discounted present worth total shows an amount of $98,745. The difference between this figure and the total (undiscounted) $164,100 in the time-line chart totals $65,355. This is the carrying charge or interest to be paid to the investors who plan to back the proposed project with money loaned at the rate of 10% interest per annum.

The raw land value must still be separated from the subdivider's profit. Anticipating a 15% profit of the raw land value, the subdivider will total the raw land and profit $98,745 and divide this figure by 115%. This obtains a raw land value of $85,865 and the subdivider's profit of $12,880. The rounded figure for the 40 raw undeveloped lots is $86,000.

The net income is negative at the beginning of the time-line and at the end of the first year, since expenses are paid out when no lots are yet developed for sale.

Methods Of Subdivision Lending

Here is a list of fees and charges to be paid by the developer. Although no attempt is made to itemize certain improvements which may be required for any one subdivision, the items common to any such list might be as follows:

Engineering fees
Surveyor's fees
Grading (on subdivision tract)
Municipal annexation and zoning fees and costs
Sanitary sewer lines (off tract)
Sanitary sewer tap-in fee

TRACT BUILDERS
- FHA/VA/FNMA Takeouts
- Interim Financing
- Joint Venture Capital
- Competitive Commitments

INVESTORS GROUP
John Jones (222) 000-0000

This advertisement is typical of syndicators active in metropolitan areas but not necessarily limiting their operations to these areas.

Figure 9-5

Sanitary sewer lines (on tract)
Storm sewer lines (off tract)
Storm sewer tap-in fee
Storm sewer lines (on tract)
Water lines (off tract)
Water tap-in fee
Water lines (on tract)
Water system (private)
Sewage system (private)
Streets
Curbs and gutters
Sidewalks
Gas lines
Telephone lines
Electricity lines

These items will vary with the area in which the subdivision is planned, but the list is fairly complete. Some are fees required by governmental units, but you will usually submit such fees out of your own funds and not out of loan proceeds. Figures 9-5 and 9-6 indicate subdivision money sources.

An LVR for financing these items would be close to 60% of

TRACT FINANCING AVAILABLE
All Projects
No Maximum
FINANCE CONSULTANTS • (213) 000-0000

This ad is typical of agencies representing specific loan correspondents interested in large block loans. The key here is "no maximum." Very likely this is from an insurance company.

Figure 9-6

the value of the raw land as appraised for present worth. In our example this would be 60% x $86,000, or $51,000. If this amount covers the first few items listed above, a new value will be established by a second appraisal designed to also cover the risks of present worth after the installation. As development costs are lessened, the value increases.

When tract sites are ready for building, releases of the individual building sites must be provided in the mortgage or the security instrument providing funds for payouts on the site improvements. Usually your lender will arrange the release price against that particular portion of the mortgage, according to value of the site, location and the like.

You can anticipate a financing schedule for tract mortgage lending to proceed in the following steps.

1. Ideally, you should hold title to the acreage. In advance of obtaining ownership, determine the highest and best use of the land through your own investigations.

2. Make up your own preliminary time-line schedule and present worth calculations before applying for the loan.

3. Expect an advance of around 60% of the raw land value as determined in your present worth calculations. After expending that amount for the preliminary improvements, a second appraisal will determine a new value with a new LVR.

This procedure will continue until all necessary development to the sites is made.

4. The initial loan obligation is released as one of perhaps two or three permanent mortgage plans is adopted. Or your "improved lots" package is sold to others for building development.

Although various loan programs throughout the nation are offered by tract lenders, you can expect some of the loan types to include provisions for longer and longer permanent mortgage terms. Young families make up such a large part of new housing developments that lenders must find ways to make their loans affordable. The key to the sale in the new development is the proper loan package geared to young buyers. The following types have been used in the past but are subject to new innovations and should be carefully scrutinized when undertaking the marketing program. These types are the *blanket mortgage,* the *variation blanket mortgage,* and the *"flexible" mortgage.*

The simplest financing found in subdivision funding methods is the *blanket mortgage* placed upon the entire tract. As the project develops and each lot is fully improved, a predetermined amount of money is disbursed for further lot development. Advances are also made as the installations are in their proper places. Progressions are made in accordance with LVR scales. This type of loan is handled as a non-home loan or as a commercial loan but carries with it the expense and trouble of refinancing when it comes time to refinance the individual houses. This is a disadvantage. In addition, the existing mortgagee may lose the long term loan opportunity to an outside lender, since permanent mortgages must be arranged later. Even so, this blanket type subdivision mortgage is used more often than other loans. In *variation blanket* mortgages, the total sum is stated in the note or mortgage but a specified portion of the total is earmarked for a separate improvement for each lot. This arrangement is not treated as a commercial loan but as a home loan and therefore carries with it a higher LVR. A separate home loan account can be created under this method covering each of the tracts.

Advances are then made as these lots are improved. This assures the lender of the permanent mortgage financing as the final structures are sold. Marketing plans seem more dependable with this plan, also. Another popular plan in recent years is the *flexible mortgage plan*, taking a separate mortgage on each tract while using a flexible contract form in the mortgage which includes an open-end provision as well as other standard benefits. This type of loan is placed on an individual lot in the name of the subdivider at a maximum LVR based upon the completed house and lot as contemplated. This plan is actually a construction loan agreement and an assumption by the final borrower of the completed house and lot. The original loan is made to the subdivider specifying how the money is to be advanced until the owner-occupant takes over. At this time the final proceeds of the loan will be disbursed. This is a big help in the final marketing of the property because the builder can tell his prospects the details of the financing and how the loan can be reduced, if additional down payments are made in cash. He can also show the prospect the advantages of avoiding the necessity of refinancing.

"Your" Regulations

Following your own rules will prevent many mistakes. Here are some suggestions for building and financing homesites which you can adopt for your own. They are safe bets, given the choices which confront you every time you begin a subdivision. Appendix XII shows a land description conversion guide for relating acreage to a sectioned land.

- Don't buy land just because it is cheap. The conveniences you will need are no doubt a long way off from this low priced "opportunity."

- Don't buy odd-shaped parcels of land to be used for conventional tract development. But unconventionally-shaped parcels may be assets in building specialty developments such as golf and tennis condos or high-priced homesites. Odd-shaped or "difficult" lots such as hillsides are worth the extra building expense and trouble and can be sold to a more affluent market, if the homesite is highly improved. Luxury housing with a view can be readily marketed despite the unconventionality of the site size and location.

- Check the development program of FHA and GI to see what opinion they offer on developing this particular land. You can depend upon unbiased free information from them.

- Don't let low taxes influence you. With low taxes go the lack of municipal services such as garbage collection, street lighting, storm sewers and police protection.

- Don't skimp on the quality of the street improvements. If the asphalt and cement start cracking up before your subdivision sites are completely marketed, prospective buyers will disappear fast.

- Check zoning adjacent to your proposed building sites. Gather adequate environmental information for the area.

- Study a topographic map for the site you are about to buy. You should have a master plan for the whole tract, even though you are plotting a smaller portion.

- Be careful of the tax burden when you carry large tracts of land for future development. Your carrying charges plus taxes can become burdensome. But you may be able to contract with landowners who allow options on long terms. This way, they pay the taxes while you shop for conditional commitments and other preliminaries.

- Take your master plan to the public utilities companies, as well as park, school, street, sewer and other city departments. If your tentative tract map is in process you will have already familiarized yourself with some of these agencies.

- Do not purchase land in an area of high costs while planning a low-cost type of development. Your marketing methods will not be in tune with the buying public. The same situation exists in low-cost areas with high-cost buyers.

- Do not offer single family homes in an area predominated by rental housing. Rental units are separate markets with separate interests. Do not have a different market with different investor interests.

- Do not try to adapt rough and hilly topography and woody areas to low cost tracts. The high cost of the land and preparation make such developments nearly impossible, and there are marketing problems.

Future drainage and run-off surface water can be a problem with adjacent land owners if no provision is made for the possibility of flooding. As the subdivision nears completion the run-off surface water increases. Plan for this by increasing site preparation, drains and channels as construction proceeds.

Modern tract planning includes the cul-de sac, a definite safety factor as well as drainage improvement. The grid street pattern used for years is actually the result of minimum planning. Offsite parking dictates adequate street width. Planning boards now usually block tract plans with narrow streets, but they are still found in older subdivisions. Streets laid out in curves can improve marketability.

Tracts proposed near the fringe of the urban area demand public sidewalks, and although planning boards may overlook this factor your lender may insist upon it as a requirement for the loan. On the other hand, sidewalks have no real significance in outlying areas where the sites are large, particularly the "junior estate" type of development, and can be eliminated in the planning. In planning for sidewalks, pay attention to the needs and trends of the area. Do not reduce the size of lots to get one or two more—the value of these "extra" lots, as well as that of some of the others, may fall and offset their extra sales potential.

Study your protective covenants and restrictions which

appear on your purchaser's deeds. These covenants and restrictions will assure the quality and success of your development if they regulate and govern all improvements to be built in the area.

Federal law has eliminated discriminatory covenants or race restrictions, and in the event you request an Environmental Review from the Housing and Urban Development you must certify with that agency that pursuant to the requirements of the regulations you will not decline to make available any part of your development to any race, color, creed or national origin. The application for such an Environmental Review is shown in Figure 9-7. The certification required to market the tract development is at the bottom of the application.

The HUD Handbook 4135.1 REV-1 describes procedures for approval of single family proposed construction applications in new subdivisions. The revision covers any pending applications for a subdivision analysis or request for an environmental review. The Revision is dated 10/2/79 and shows two procedures - Developers Certification and Local Area Certification. When a local area study has already resulted in a Local Area Certification, no future applications for subdivision analysis or an environmental review is to be requested by developers. Individual or group applications for a developer's single family proposed construction in a new subdivision may proceed directly with the Developer Certification steps.

Developer Submits, with Exhibits:

1. Application for Environmental Review (Form HUD 92250). Figure 9-7.

2. Equal Employment Opportunity Certification (Form HUD 92010).

3. Location Map.

4. Preliminary Subdivision Sketch Plan.

Form Approved
OMB No. 63R-1203

U.S. DEPARTMENT OF HOUSING AND URBAN DEVELOPMENT
HOUSING - FEDERAL HOUSING COMMISSIONER

APPLICATION FOR ENVIRONMENTAL REVIEW

NAME AND ADDRESS OF DEVELOPER

CITY OR COUNTY	STATE	ZIP CODE	TELEPHONE NUMBER *(Include Area Code)*

HUD FILE NUMBER	NAME OF SUBDIVISION
TRACT NUMBER	LOCATION OF SUBDIVISION

Environmental review of this proposal is requested and the following exhibits are attached:

- ☐ Location Map
- ☐ Preliminary Subdivision Plan
- ☐ Signed Equal Employment Opportunity Certification HUD-92010
- ☐ Affirmative Marketing Plan *(HUD-935-2/Certification of Intent Not to Market (HUD-935-3))*

Optional Information submitted:

- ☐ Soils Report
- ☐ Preliminary Grading Plan
- ☐ Other ____________
- ☐ Topographic Data
- ☐ A-95 Comments

GENERAL INFORMATION

1. Developer is: ☐ Land Owner ☐ Option Holder
2. Size of this Parcel: ____________
3. Number of Lots: ____________
4. Typical Lot Size: ____________
5. Adjacent Land Under Developer's Control: ____________ Acres
6. Other Land Uses ____________
7. Parcel is part of a locally-approved development plan of _____ lots
8. Developer will:
   ☐ develop land and build homes; initial plan is to start _____ homes in $ ____________ to $ ____________ price range.
   ☐ develop land and sell improved lots.
   ☐ produce finished typical building sites at a price of $ ____________.
   ☐ limit site grading to minor redistribution with house foundation in natural soil.
   ☐ construct house foundations on soils engineered cuts and fills.
9. Any Special Assessments? ☐ Yes ☐ No
   If yes, describe under Remarks.
10. Any Mineral Reservations? ☐ Yes ☐ No
    If Yes, describe under Remarks.
11. Bus *(distance, direction, frequency)*: ____________
12. Fire Station *(distance, direction)*: ____________
13. Shopping *(distance, direction)*: ____________
14. Is Tentative Map approved by Local Authorities? ☐ Yes ☐ No
15. Is Plan recorded? ☐ Yes ☐ No
16. Covenants recorded? ☐ Yes ☐ No
    ☐ Will be identical to previous unit.
    ☐ Will conform with FHA Data Sheet 40.
17. a. Water System: ☐ Central *(Public or Community)* ☐ Individual
    b. Sewerage System: ☐ Central *(Public or Community)* ☐ Individual
18. Proposed Street Improvements:
    Pavement Base ____________
    Wearing Surface ____________
    ☐ Curb and Gutter ☐ Sidewalks
19. Underground electric and telephone? ☐ Yes ☐ No
    (Explain under Remarks)
20. Gas? ☐ Yes ☐ No
21. Will development include common area? ☐ Yes ☐ No
    (Describe facilities and maintenance under Remarks)
22. Developer ☐ has ☐ has not previously dealt with this HUD Office. ☐ has dealt with other HUD Offices.
    (Explain under Remarks).
23. Schools *(Distance, direction)*:
    Elementary ____________
    Junior High ____________
    High ____________
24. Historic/Archeological sites within one mile.
    ☐ Yes *(Attach Description)*
    ☐ No
25. ☐ Planned Unit Development:*
    Units____1 Br___2Br___3Br___4Br ___Total ________
    Off-Street Parking Spaces ____________
    Planned Common Areas ____________

**Any subdivision where there is a mandatory membership in a Homeowner's Association requiring lien supported assessments will be processed as a Planned Unit Development (PUD).*

REMARKS *(Use additional sheet(s) if necessary)*

CERTIFICATE

By submitting this request for site eligibility and signing this certification, builder, developer, seller or other signatory agrees with the Federal Housing Commissioner that pursuant to the requirements of the HUD Regulations, (a) neither it nor anyone authorized to act for it will decline to sell, rent or otherwise make available any of the properties or housing in the subdivision to a prospective purchaser or tenant because of race, color, creed or national origin; (b) it will comply with state and local laws and ordinances prohibiting discrimination; and (c) failure or refusal to comply with the requirements of either (a) or (b) shall be a proper basis for the Commissioner to reject requests for future business with which the sponsor is identified or to take other corrective action he may deem necessary to carry out the requirements of the Regulations.

The undersigned further agrees, in consideration of the review of this proposal by HUD, that any deposit or downpayment received by the undersigned or an agent of the undersigned in connection with the purchase of a home within the subdivision described above shall, upon receipt, be deposited in escrow or in trust or in a special account which is not subject to the claims of my creditors and where it will be maintained until it has been disbursed for the benefit of the purchaser or otherwise disposed of in accordance with the terms of the contract of sale.

SIGNATURE OF DEVELOPER	TITLE	NAME OF DEVELOPMENT COMPANY	DATE

Replaces Form FHA-2250, which is Obsolete

HUD-92250 (6-79)
VA-26-8492

Application for Environmental Review
Figure 9-7

DEPARTMENT OF HOUSING AND URBAN DEVELOPMENT
FEDERAL HOUSING ADMINISTRATION

____
(Date)

IN REPLY REFER TO:

RE: ENVIRONMENTAL REVIEW
Subdivision
Name: _______________

Location: _______________

HUD File No.: _______________

Dear

Our analysis of your application for environmental review shows that your project is environmentally acceptable. The Veterans Administration agrees.

If you intend to proceed with your development, please prepare and submit a complete certified pre-construction exhibit package as requested on the attached check list Form HUD 92256. This package shall include certification from each professional that exhibits prepared by him/her meet HUD standards. (Copy of certification form is attached.)

If your development involves site grading under fill conditions, you must certify that your final neighborhood grading and drainage plans comply with the requirements of Data Sheet 79G and applicable local criteria and standards.

Applications for mortgage insurance may be submitted after HUD acceptance of certification of pre-construction exhibits and developer's concurrence on Form HUD 92258.

Sincerely,

DHD/SOS

Attachments

Letter of Environmental Review
Figure 9-8

CONSTRUCTION
LOANS
Call: John Jones, Senior Vice President

ANY CITY BANK
7 Chestnut Avenue
Metroville, USA / (000) 111-2222

Member FDIC Equal Opportunity Lender

A direct commitment from a commercial bank assures two party lending—borrower and lender, no middlemen.

Figure 9-9

NEW CONSTRUCTION LOANS
and take - out commitments
Eagle Mortgage (000) 111-2222

This indicates an agency offering loan services which may call for special needs.

Figure 9-10

5. Evidence that A-95 clearing houses have been contacted. The Office of Management and Budget Circular A-95 is available at HUD field offices. OMB will receive a copy of your application for environmental review and location map, or, to save time, the developer may send copies of the application direct to the local or regional clearing house listed in OMB Directory.

6. Affirmative Marketing Plan. (Form HUD 935.2).

The HUD Valuation Staff will conduct the environmental review and will notify the developer whether or not the proposal is acceptable, conditional, or if subject to an Environmental Impact Statement. If your project is environmentally acceptable The Environmental Review Letter (Form HUD 92255) will be sent to the applicant. See Figure 9-8.

Local Area Certification (LAC) procedures may be found in Handbook 4135.1 REV-1 or in *Builder's Guide to Government Loans* published by Craftsman Book Company, and covering HUD insured loans as well as other government housing programs.

Lending associations or agencies offering loan services have government program information as well as conventional loan information at their disposal. Figures 9-9 and 9-10 show typical loan sources for developed lots ready for building.

10

Leasehold Financing

Leasehold mortgage financing is extremely important to any discussion of housing construction. In real property financing in many states a ground lease or a leasehold as the "real estate security" is just as important to a lending organization as is the ownership of the land. From a lender's viewpoint, if a long term leasehold interest is on the land only, then actual ownership exists in the improvements constructed upon the land. Housing costs can be minimized by leasehold financing, thereby enabling both builders and purchasers to take advantage of the savings. Marketing and promotional efforts taken by a builder who has leaseholds at his disposal will stimulate his sales—the payments are more convenient to his purchaser's pocketbooks. Leaseholds should be popular during highly inflationary times. We can be assured that the costs of shelter will continue to accelerate and in the opinion of the author this may be as much as 15% per year. By 1985 the national median price of a new home will be at least as much as $125,000 for the single-family category of construction plus land. Since lenders and new buyers alike are looking for new methods of financing, builders cannot afford to overlook leaseholds as a security instrument with marketing appeal.

Leasehold Lenders

Ground rents as a method of financing housing construction is not new, but long term land leases could hold the key to marketing problems in lending on new construction projects with their permanent mortgage provisions. Long term land leases carry a simple method of assignment. As security instruments they are readily marketable; however, some states—among them Massachusetts—have the established "fee simple" ownership requirement that may preclude leasehold lending. In many areas such as Pennsylvania and Maryland, where land is very high, leasehold loans have been applied for many years. Depending upon the state in which you conduct your operations, there is no reason why a lending institution or syndicated group could not be found that would enter into leasehold financing. Be sure all the legal conditions can be met for such an arrangement before you begin your search for a leasehold mortgage lender. Leaseholds with options to buy are a feature with some landowners.

Federally chartered savings and loans are specifically authorized to make mortgage loans on leasehold security. State chartered savings associations, on the other hand will look to their individual state regulations to determine if they have the authority to make these loans. Federal charters require leases to extend 10 years beyond the loan period.

In states that limit the power to make leasehold loans, or where the legality to make these loans may be questionable, some lending institutions may be authorized to make FHA insured leasehold loans in lieu of conventional leasehold loans. This involves a first mortgage on real estate on a leasehold (1) under a lease having a period of not less than 99 years which is renewable, or (2) under a lease having a period of not less than 50 years to run from the date the mortgage was executed. The Veterans Administration guarantees leasehold loans if the term extends 14 years beyond the loan's maturity date.

In those states that authorize savings associations to make government insured loans, the leasehold lending power is spelled out in the state statutes. These states are West Virginia, Oregon, North Dakota, Mississippi, Kentucky and Indiana. Other states leave out any mention of leaseholds as such, but use broad language which could render this security instrument acceptable under FHA. These states are Arkansas, Florida, Connecticut, Idaho, Maine, Louisiana, Michigan, Ohio, New Hampshire and Tennessee. California requires the lease to run ten years beyond the maturity of the loan. The states that specify ''real estate'' security for their lending associations' conventional mortgages do so for any FHA loans as well. Therefore it is assumed that this include leaseholds in an interpretation of real estate security. Massachusetts specifies in its cooperative bank statute that fee simple security must be the instrument used for FHA, while using the broader term of real estate security for conventional loans. This may bar leasehold lending in that state, at least under FHA. Check with a loan officer on the staff of a savings institution in your state.

Developments on leaseholds are also financed by banks, trust companies and insurance companies, particularly in large subdivisions. FHA and VA also have been active in this field, covering the insured or guaranteed loan made by various lending institutions.

Leasehold Benefits

In past years leasehold financing activities have been mostly concentrated in the field of long term commercial leaseholds in various parts of the country where the money was used for construction. A few older communities in the east used a limited amount of residential leasehold financing. While a simplified case of leasehold financing would be the prospective builder who wishes to construct one single family home on a suburban vacant lot, this is actually rare. But large housing tracts have been financed in this way. The landowner may not wish to sell the land but is willing to give the speculative builder a long term lease. If this proposal is made in a state permitting leasehold lending, the builder can get construction money based upon the value of the leasehold plus the improvements. If a state charter outlaws leasehold financing, the borrower-builder may be able to obtain a leasehold loan from a federally chartered savings group in any state in which the word ''federal'' must appear in the lender's name if it is a federally chartered association.

There are several reasons why leasehold loans are favored, besides the obvious price problem in the housing market. Here are four common conditions calling for the leasehold alternative.

1. Rare sites with extremely high popular demand, as in ocean-front or lakeside property.

2. A scarce supply of land which can be bought in fee simple estate.

3. A high ratio of land cost to improvement cost.

4. Estate taxes and income tax considerations.

Since young families are being priced out of the market, this type of financing is far overdue. If builders can lease a plot of land with assignment or an option to buy, this long term ground lease can be used for subdividing and improving the lots to obtain construction loans for building houses, just as in any other type of financing. As you sell each house, your buyers will be assigned the ground lease for the lot underlying that house. The lender can grant the construction loan to you, the builder, for the erection of the houses. The purchase of each house can be financed by an assumption of the construction loan/take-out, or a new permanent mortgage loan can be granted.

The best protection for all concerned is an *early lease plan.* The plan should be set out in the lease indicating a protection for the lessor (owner of land), the lessee (mortgagor), and the mortgagee (lender). The negotiations between lessee and the lessor progress more satisfactorily if the mortgage lender is involved early in the negotiations. The lease should be drafted so that the leasehold interest is to be subsequently mortgaged under specific terms and conditions. These terms and conditions can be written into the lease in the beginning. Under an early lease plan, any covenants or conditions which may be objectionable to the lender may be eliminated through a possible subordination agreement with the lessor, or landowner. Your lender will probably insist on the landowner subordinating all privileges under the lease except that of the payments of actual ground rents or any charges considered to be a charge on the property. Of course, any covenants that cannot or will not be removed from the title to the land are closely examined by your lender. If any covenants appear detrimental to the lender, he simply will not make the loan until the change is made. Restrictive covenants and a continuing repair and maintenance obligation provided in the leasehold conditions will forestall depreciation.

In areas where this method of financing has been used for many years, lenders do not hesitate to lend money for construction based on leaseholds. Inasmuch as the ground rent contracts are assignable in case a mortgage or trust deed on the building is eventually foreclosed, the lender can assume the borrower's rights and obligations under the ground lease. The lender then has a property that he must dispose of subject to the rent from the leasehold. It then makes little difference to him whether he finances a construction project where the borrower has full title ownership of the site but needs to borrow most or all of the cost of building, or where the borrower does not own the land but needs to borrow a smaller proportion of the construction cost. The lender will look at the risk rates of two separate systems. For instance, a proposed project fully completed will be worth $57,500 of which lot value is $14,375 and structure is $43,125. The purchaser may obtain the property for 25% down payment. Ownership of the lot free and clear of encumbrances will then be qualified for a 100% loan on the structure. Or it would be equivalent to a 75% loan to value ratio for construction, subject to a leasehold in the amount of $14,375. The lender's risk equivalents are as follows:

Ownership in fee title		Leasehold contract	
100% construction loan	$43,125	75% LVR as fully improved	$43,125
Cash paid for land 25%	14,375	25% leasehold interests	14,375
Property value	$57,500	Property value	$57,500

The 75% loan to value ratio is applied to the full $57,500 in the case of a leasehold mortgage, in order to illustrate the equivalency.

The "Buffer Zone"

The lender will assure himself that the term in years is long enough to protect his interests; in some states he can make a loan only on leaseholds extending a specific number of years after the actual loan term has expired. The federal savings associations in any state must follow the 10-year pattern—that is, the lease must extend or be renewable automatically for a period of 10 years after the terminal date of the leasehold mortgage loan. There are no minimums set in some states, so the lender will specify an extension of the lease for an adequate length of time beyond the maturity of the loan, or the lender will specify this power to enforce renewal of the lease for as long as he has the loan outstanding on the leasehold. Leasehold safety depends on the period of the lease, or upon an extension of the lease beyond the length of the loan and a low risk in case of default on the payments. The lender will exercise the renewal option at the time of the original closing of the loan which guarantees his right to extend the lease at any time during the term of the loan. This is a lender's "buffer zone." If your lending source is a commercial bank which operates under national banking regulations, remember that his provisions call for a ten year extension past the mortgage loan period.

Marketing The Leasehold Bundle

A lease is property ownership in the sense that it must provide for your unlimited right of assignment to your successors, if it is to be accepted as mortgage security. A simplified explanation is in order for the builder, investor, speculator, subdivider or developer who operates in the housing field with leaseholds. It is extremely important that you, as lessee, have a right to mortgage the leasehold free of any covenants that would interfere with assignment to others. And this characteristic must remain a permanent part of the lease for you to sell it to subsequent purchasers. Without this definite provision, the leasehold interest would be unmarketable in eventual foreclosure or assignment in lieu of foreclosure. Future lessees (purchasers) must be free to assign their interests in the property without obtaining the lessor's (landowner's) consent. Keep in mind that the mortgage loan is against the building structure only in the repayment of the loan, but that the lease, along with the building, is taken over upon mortgage foreclosure. Further provision should be included in the leasehold stipulating that the assigned liability ends at the time the lease is assigned. An assignee may be liable for obligations which have accrued while he holds the lease, but not for those obligations that accrue after he assigns the lease to another purchaser.

In localities where ground rents are often bought and sold, the contracts are freely traded as are mortgages or trust deeds in other areas. A market is established for this basic underlying lien for the investment of trust funds or similar programs of sound investment. Building operations can gain additional profit from selling, on a capitalized net income basis, the ground rents which were originally purchased at a much lower price. Speculative builders can use leaseholds effectively. They realize that the value of the ground rent is determined by the income it produces, and a vacant parcel of land is not easily marketed because its income is little or nothing. When a suitable building or structure is erected on the parcel or site, the contract for the ground rent supplies income for capitalized value. The tenant usually pays all taxes and other charges while the income from ground rent is net return. Net ground leases for new developments, particularly in California and Hawaii, have been used to an ever-increasing extent for single family dwellings. In Hawaii, leasehold lending is a common custom. Kaiser Industries make exclusive annual ground rents in their Hawaii-Kai development. Ground rents are usually paid annually at a rate that is a percentage of the value of the lot established in the original appraisal.

Be familiar with the terms *contract rent* and *economic rent* in leasehold financing, because the tenant's permanent ground rent payments remain the same under contract rent regardless of inflationary factors affecting economic rent. Contract rent doesn't change, and the terms of the contract indicate when any renegotiations are due. The Irvine Company of Southern California has many thousands of acres under long term ground lease on property in the single family category reaching well over the $200,000 valuation. Leasehold rents to the Irvine Company are sometimes payable semi-annually and sometimes quarterly directly to the landowner by the homeowner.

Where leasehold financing is fairly common, the owner of the leased land does not do the subdividing or construction of houses. Instead, a *master lease* is created for the subdivider who makes such improvements as streets and sewers, grading and plot preparation. The subdivider then either constructs homes on the lots or sells the entire "package" to a builder. The cost of the improvements as installed by the subdivider is added on to the price of the homes with assignment of a lease to the developer. Sometimes the lease terms will carry specific hedges against inflationary conditions, requiring a renegotiation 25 or 30 years hence. An example is the typical ground lease shown at the end of this chapter, in which ground rents must be adjusted after 30 years to offset the changes in the cost-of-living scales at the end of the lease period. Other leasehold transactions may call for a full reappraisal of the land at the end of, say, 25 years, and a new percentage ratio applied to the new land value. This could change from a former 6% return, or whatever a fair return in the ground rent would be at the time of reappraisal, to an 8% return per year. An owner of the home is given credit for his leasehold interest in addition to a regular appraised value of the property in determining its worth for the new adjustment. In the past a ratio of 6% ground rent to the full appraised value of the land was considered fair. The market for ground rent stabilizes by itself, just as supply and demand do in sales prices of homes or other commodities. Landowners are known to be conservative in their ground rents. Not only do they not want to get the reputation of "holding up" the market in homes but it is to their advantage to keep long term tenants satisfied. Leaseholding for income would defeat its purpose if there were defaults due to the high rent. Lessors would have the extra expense of management problems with property dormant due to defaults.

In terms of defaults, you as a speculative builder, subdivider or participant in leaseholds should regard a lease as an ownership like any other property. It can be sold, discounted, borrowed upon, foreclosed or handled like any mortgage security. Therefore, a default would be handled exactly as a mortgage or trust deed foreclosure; the lender should make sure he has a protection clause inserted that will assure him of such notice by the landowner or lessor of a default in ground rent. In other words, since the landowner has no investment in the building improvements he is not responsible or particularly concerned about his tenants' payments to other debtors—only his own rental income. A provision in the lease must provide that the notice of default be served simultaneously from the landowner to the lender as well as from the landowner to the tenants. The lender is protected if the tenant's default can be taken over by the lender until a proper disposition can be made. Therefore, obtaining the current status of ground rent information is important to the lender. Actual bankruptcy on the part of the lessee resulting in default in leaseholds cannot be cured by the lender, even if notice is served that bankruptcy exists. So the lender checks out the lease thoroughly to see that no incurable defaults are likely to arise. This is the feature that separates the leasehold from the fee simple estate which makes leasehold unacceptable as mortgage security by some state legislatures and by lenders who are more sensitive to risk than others.

Fee Title and Leasehold Value Equivalents

Builders who anticipate construction financing for projects on leased land or land that will be negotiated for a lease must "think" like the lender's mortgageable lease. Ask yourself these questions as if you were a leasehold lender:

1. Can the lease be negotiated to extend beyond the permanent loan maturity date, whether or not my state's laws provide a minimum? How far beyond?

2. Can a subordination agreement be made transferring all rights, privileges and remedies accruing to the lessor (landowner) under the proposed lease—*except* ground rents—to me?

3. Do any covenants, conditions, or reservations exist now that I cannot accept without its exclusion in the subordination agreement?

4. Can I obtain the unrestricted right of assignment to me and my successors without a covenant against assignment to subsequent buyers, and without the lessor's (landowner's) consent?

The last question relates to foreclosure, or taking over the lease by assignment in lieu of foreclosure, because the leasehold interests must be marketed by you if you were the lender. If you cannot market the interests you cannot realize on its security in case of default. Once the mortgageable ground lease is established, it can be marketed as follows:

A young family cannot afford more than $8,625 for a modest house and lot at $57,500. A sale price of $43,125 with a 20% down payment ($8,625) would appear more attractive. Here are the buyer's value equivalents.

FEE TITLE		LEASEHOLD	
Price	$57,500	Price	$43,125
Down payment	8,625	Down payment	8,625 (bldg.)
Loan of 80% LVR	46,000 land & bldg.	Loan of 80% LVR	34,500 (bldg.)
2nd mortgage/TD	2,875	Market value	14,375 land
		(before adjustment for ground rent)	

Here are the differences in monthly payments between the two types of ownership.

FEE TITLE		LEASEHOLD (land value reduced 30%)	
$46,000 @9½% 25yrs.	$401.91	$34,500 @9½% 25 yrs.	$301.45
$ 2,875 @10% due 3 yrs	28.75(balloon)	$10,062 @7½% G. rent	62.91
Total	$430.66	Total	$364.36

Consider the differences in each kind of ownership. The ownership in fee title carries the large balance due on the balloon payment at some future date. This junior lien is actually a continuation of the down payment which is in the form of credit. The lot cost is a part of the overall price; therefore, much more mortgage money is required reflecting more interest dollars to be spent on both the 1st and 2nd mortgage/trust deeds. On the other hand, even with low down payments leaseholds have always been conservative with their ground rents, compared to the usual prevailing interest rates on mortgages. Ground leases written for long terms have the percentage factor specified in advance to any future renegotiations. The typical landowner uses the appraised value of the raw land as a basis for applying a fair rate of return. In some areas the landowner has taken an appraised value and cut it back by as much as 25% to 35%. To be competitive with prevailing interest rates, he usually figures his rental on his reduced valuation to be considerably lower than mortgages with government insured or guaranteed loan rates. So it is an advantage to him to retain long term tenants with new terms to be negotiated a few years ahead. Since there is a disadvantage to the purchaser of property based on leasehold—a lack of fee title to the land—contract rent is more attractive than economic rent. Savings can be substantial when compared to economic rent in an area where land values continue to rise as your contract rent remains constant. In theory, at least, if landowners leased their properties based on "fair market value" of the full fee title—without adjustment—buyers would seek comparable prices elsewhere. This assumes that the land is "bought" as a whole package. Thus landowners may well ask themselves the question, "Why will buyers prefer my lease contract over full title ownership?" The answer is, because they can't buy identical property with the same use at the same or with comparable budget payments.

Established ground rents of the past have always carried the provision for a periodic reappraisal of the land. It is these reappraisals or rent adjustments that establish the base generally at a percentage of the land value after a generous reduction. For many years this has been 4% - 6% of the reduced fair market value. Isolated cases have been known to be as low as 2½% per year for ground rent, or contract rent. Savings realized during the years when interest rates are constantly rising can be substantial, even with periodic adjustments.

Here is the concept of contract rent as opposed to economic rent presented as a comparison of the two on one piece of land. We can use a lot that would have been appraised at fair market value (for sale in the open market) at $14,375. The lot owner cuts back the value 30% to $10,062.50. He then selects a rental figure at, say, 7½% per year on the reduced evaluation. Our comparison uses a value equivalent of $57,500.

ECONOMIC RENT						
Percent return	6%	6.75%	7.50%	8%	9%	9.75%
Annual rent	$604.	$679.	$755.	$805.	$906.	$981.

CONTRACT RENT						
Ground rent	-0-	-0-	$755.	$755.	$755.	$755.
Annual bonus	-0-	-0-	-0-	50.	151.	226.

Ground rents are usually paid annually but have been calculated on a basis of 12 months in the leasehold for monthly payments.

A ground lease typical of subdivided land with a landowner's general covenants and provisions is shown below with supplemental documents. The supplemental agreements provide permission for a lender to assign under the mortgage/trust deed and must be made part of the ground lease. The Consent to Assignment of Lease specifies the amount of the "note" to be secured by the leasehold, providing for notice to lenders upon default of the rental, and includes a separate Acceptance and Agreement.

Here are the circumstances of this ground lease. A landowner, or lessor, makes a lease deal with a subdivider, or lessee, for a 60 year period. There are terms with general covenants and conditions including provision for assignment. Also included are specific instructions to obtain permission for any loan encumbrances on the leasehold. Provision for adjustments to the terms of rental in the future is also included. In order to obtain a loan to construct improvements the lessee, subdivider, tenant or any subsequent sub-lessee accompanies his chosen lender to the landowner for permission to mortgage the ground lease. If the landowner is in agreement, he executes the Consent to Assignment of Lease by Mortgage (or trust deed) specifying the mortgage amount, the lender's name and the date. The lender (mortgagee) or assignee in turn, joins with the subdivider, lessee, tenant or assignor in executing the Acceptance and Agreement as a part of the Consent. Ideally the three documents, Lease, Consent and Acceptance/Agreement should be notorized and recorded together.

Note that in article no. 23 of the Ground Lease the landowner has protected himself by omitting the method to be used in adjusting rental in 30 years. Only the basic wording of "then fair market value" will provide the value figure the 7½% is to be applied to. In article no. 6, improvements are specified in detail for further protection of lessor, or landowner. The Lease provides for the tenant to sub-lease, and consent shall be withheld by the lessor until defaults or other offenses are satisfied under the Lease. This is found in article no. 12. Although the possibility may be remote in the future, disagreements as to fair market value can be handled under article no. 23 of the Ground Lease, by arbitration.

WHEN RECORDED MAIL TO:

LANDGUIDE ENTERPRISES
ADDRESS:
ALL METROS, STATE, USA, ZIP

Attn: Builder & Residential Lease Management
Residential Division

Space above for Recorder's use only

LEASE WITH OPTION TO PURCHASE

TRACT NO. _____

THIS LEASE, made this first day of_____________________, by and between LANDGUIDE TRUST CORPORATION, a State Chartered corporation, as Trustee under Trust No. 14S-7211-02, herein referred to as "Lessor," whose address is c/o Landguide Enterprises, All Metros, USA ZIP
and ____
____
____
herein referred to as "Lessee", whose address is ______________________
____.

W I T N E S S E T H:

1. PROPERTY LEASED: For and in consideration of the payment of the rents and taxes and other charges, and for the performance of all of the covenants and conditions of this Lease by Lessee, Lessor hereby leases to Lessee that certain lot of land situated in the City of ____________________________, County of______, State of _________, described as follows, to-wit:

RESERVING UNTO LESSOR, its successors and assigns, items reserved to Landguide Enterprises, a state chartered corporation, as Grantor, as described in Corporation Grant Deed attached hereto as Exhibit A, which reservations are hereby incorporated herein by reference.

SUBJECT TO oil, gas, current taxes, assessments, covenants, conditions, restrictions, reservations, and all other items more particularly described on Corporation Grant Deed attached hereto as Exhibit A, which items are hereby incorporated herein by reference.

(hereinafter referred to as "leased land").

2. TERM OF LEASE: Said land is leased for a term commencing on______ __________________and ending on______________________________, subject, however, to earlier termination as hereinafter provided.

28/1/11-30-77

3. RENTAL: Lessee agrees to pay to Lessor as rental for the use and occupancy of said leased land during the term of this Lease the sum of_________ ____ Dollars ($____________) per year, in advance, on the first day of__________ of each year of said term, beginning on_______________________________; subject, however, to adjustment at the time and in the manner as herein provided for in the Article entitled "RENTAL ADJUSTMENT". Any installment of rent accruing under the provisions of this Lease which shall not be paid when due shall be subject to a late charge of Twenty-Five Dollars ($25.00) plus one percent (1%) per month from the date when due and payable by the terms of this Lease until the same shall be paid. All rentals hereunder and charges with respect thereto shall be paid in lawful money of the United States.

4. TAXES AND ASSESSMENTS: In addition to the rents above provided, Lessee shall pay, prior to the delinquency date thereof, all taxes and general and special assessments of every description which during the term of this Lease may be levied upon or assessed against the leased land and all interest therein and improvements and other property thereon, whether belonging to Lessor or Lessee; and Lessee agrees to protect and hold harmless the Lessor and the leased land and all interest therein and improvements thereon from any and all such taxes and assessments, including any interest, penalties and other expenses which may be thereby imposed, and from any lien therefor or sale or other proceedings to enforce payment thereof.

5. USE OF LEASED LAND: Lessee shall use the leased land solely for private single family residential purposes, and Lessee shall not construct or maintain thereon or permit to be constructed or maintained thereon more than one single family dwelling, and Lessee shall not use or permit any person to so use the leased land and the improvements thereon, or any portion thereof, as to disturb the neighborhood or occupants of adjoining property, or to constitute a nuisance, or to violate any public law, ordinance or regulation from time to time applicable thereto.

6. IMPROVEMENTS: No structure (including but not limited to, fencing of any type) or material addition to or alteration of the exterior of any building constructed on the leased land, shall be commenced unless and until plans and specifications covering the proposed structure, addition or alteration shall have been first submitted to and approved by THE LANDGUIDE TRUST CORPORATION. When any construction is commenced on the leased land, the same shall be prosecuted with reasonable diligence until completed, and shall conform to all public laws, ordinances and regulations applicable thereto, and shall be constructed and completed at the sole cost and expense of Lessee and without any cost, expense or liability of Lessor whatsoever.

7. MAINTENANCE OF LEASED LAND: Lessor shall not be obligated to make any repairs, alterations, additions or improvements in or to or upon or adjoining the leased land or any structure or other improvement that may be constructed or installed thereon, but Lessee shall, at all times during the full term of this Lease and at its sole cost and expense, keep and maintain all buildings, structures and other improvements on the leased land, if any, in good order and repair, and the whole of the leased land and all improvements thereto free of weeds, and rubbish, and in a clean, sanitary and neat condition, and Lessee shall construct, maintain and repair all facilities and other improvements which may be required at any time by law upon or adjoining or in connection with or for the use of the leased land or any part thereof, and Lessee shall make any and all additions to or alterations in any buildings and structures on said premises which may be required by and shall otherwise observe and comply with any and all public laws, ordinances and regulations for the time being applicable to the leased land, and Lessee agrees to indemnify and save harmless the Lessor against all actions, claims and damage by reason of Lessee's failure to keep and maintain said premises and any buildings and improvements thereon as hereinabove provided, or by reason of its non-observance or nonperformance of any law, ordinance and regulation applicable thereto.

8. <u>RESTORATION OF IMPROVEMENTS</u>: If, during the term hereof, the dwelling, structures or other improvements, if any, constructed by or for Lessee on this leased land, or any part thereof, shall be damaged or destroyed by fire or other casualty, Lessee may, at its cost and expense, either (a) repair or restore said dwelling and improvements in accordance with plans and specifications approved in accordance with the Article hereof entitled "IMPROVEMENTS", commencing such repair or restoration within one hundred fifty (150) days after the damage occurs and thereafter pursuing said repair or restoration to completion with due diligence, or (b) subject to the consent of an encumbrancer, if any, tear down and remove the same from the leased land. If Lessee shall elect not to repair or restore said damaged dwelling, Lessee shall, within one hundred fifty (150) days after said damage occurs, tear down and remove all parts thereof then remaining and the debris resulting from said fire or other casualty, and otherwise clean up said premises as herein provided for in the Article entitled "REMOVAL", and any failure by Lessee so to do shall constitute a breach of the covenants and conditions of this Lease. If the Lessee shall elect not to repair or restore as aforesaid, this Lease shall cease and terminate one hundred fifty (150) days after the date on which said damage occurs.

9. <u>LIENS AND CLAIMS</u>: Lessee shall not suffer or permit to be enforced against Lessor's title to the leased land, or any part thereof, any lien, claim or demand arising from any work of construction, repair, restoration, maintenance, or removal as herein provided, or otherwise arising, except liens, claims or demands suffered by or arising from the actions of Lessor, and Lessee shall pay all such liens, claims and demands before any action is brought to enforce the same against said land; and Lessee agrees to hold Lessor and the leased land free and harmless from all liability for any and all such liens, claims or demands, together with all costs and expenses, including but not limited to, reasonable attorneys' fees and court costs incurred by Lessor in connection therewith. Lessor shall have the right at any time to post and maintain on the leased land such notices as may be necessary to protect Lessor against liability for all such liens or otherwise. Notwithstanding anything to the contrary contained in this Article, if Lessee shall in good faith contest the validity of any such lien, claim or demand, the Lessee shall, at its expense, defend itself and Lessor against the same and shall pay and satisfy any adverse judgment that may be rendered thereon before the enforcement thereof against Lessor or the leased land, and if Lessor shall require, Lessee shall furnish to Lessor a surety bond satisfactory to Lessor in an amount equal to such contested lien, claim or demand indemnifying Lessor against liability for same, or if Lessor shall request, Lessee shall procure and record the bond provided for in Section 0000.1 of the applicable state Code of Civil Procedure, or any comparable statute hereafter enacted providing for a bond freeing the leased land from the effect of such lien or claim or action thereon.

10. <u>LIABILITIES</u>: Lessor shall not be liable for any loss, damage or injury of any kind whatsoever to the person or property of Lessee, or any of Lessee's employees, guests or invitees or of any other person whomsoever, caused by any use of the leased land, or by any defect in any building, structure or other improvement constructed thereon, or arising from any accident on the leased land or any fire or other casualty thereon, or occasioned by the failure on the part of Lessee to maintain said premises in safe condition, or by any nuisance made or suffered on the leased land, or any improvements thereto, or by any act or omission of Lessee, or of any member of Lessee's family or of Lessee's employees, guests or invitees, or arising from any other cause whatsoever; and Lessee hereby waives on its behalf all claims and demands against Lessor for any such loss, damage or injury of Lessee, and hereby agrees to indemnify and save Lessor free and harmless from liability for any such loss, damage or injury of other persons, and from all costs, expenses and other charges arising therefrom and in connection therewith.

11. LESSOR PAYING CLAIMS: Should Lessee fail or refuse to pay any tax, assessment or other charge upon the leased premises when due and payable as provided herein, or any lien or claim arising out of the construction, repair, restoration, maintenance and use of the leased land and the buildings and improvements thereon, or any other claim, charge or demand which Lessee has agreed to pay under the covenants of this Lease, and if after thirty (30) days written notice from Lessor to Lessee and to its authorized encumbrancer, if any, Lessee or its said encumbrancer shall fail or refuse to pay and discharge the same, then Lessor may, at its option, pay such tax, assessment, lien, claim, charge or demand, or settle or discharge any action therefor or judgment thereon, and all costs, expenses and other sums incurred or paid by Lessor in connection therewith shall be repaid to Lessor by Lessee upon written demand, together with interest thereon at the rate of ten percent (10%) per annum from the date of payment until repaid, and any default in such repayment shall constitute a breach of the covenants and conditions of this Lease.

12. ASSIGNMENT: Lessee shall have no right to assign this Lease without the prior written consent of Lessor which may be obtained as hereinafter set forth; provided, however, this Lease or any right hereunder shall in no case be assigned separate and apart from the improvements located on the leased land. Lessor will consent to the assignment of this Lease as a whole provided that (a) Lessee or any of its successors or assigns shall not be in default hereunder at the time of a proposed assignment; (b) the proposed transferee shall covenant in writing with Lessor to keep, perform and be bound by each and all of the covenants and conditions of this Lease herein provided to be kept and performed by Lessee; (c) the Lessee or proposed transferee shall furnish Lessor with an executed copy of such Assignment or other document to be used to effect such transfer, the address of the proposed transferee, and the proposed effective date thereof; and (d) the transferor or proposed transferee shall pay to Lessor a transfer fee of______________________.

13. ENCUMBRANCES: Lessee may assign Lessee's interest in this Lease and the leased land to a trustee under a Deed of Trust, (or with a mortgage lien), for the benefit of a lender (herein called "encumbrancer") upon and subject to written consent of Lessor with following covenants and conditions:

A. Said Trust Deed/Mortgage Assignment and all rights acquired thereunder shall be subject to each and all of the covenants, conditions and restrictions set forth in this Lease and to all rights and interests of the Lessor hereunder; and in the event of any conflict between the provisions of this Lease and the provisions of any such Trust Deed or Mortgages, the provisions of this Lease shall control.

B. Any encumbrancer as a transferee under the provisions of this Article shall be liable to perform the obligations of the Lessee under this Lease only so long as such encumbrancer holds title to the leasehold. Any subsequent transfer of the leasehold hereunder shall be subject to the conditions relating thereto as herein set forth in the Article entitled "ASSIGNMENT".

C. Upon and immediately after the recording of the Trust Deed or Mortgage Assignment covering the leased land, Lessee, at Lessee's expense, shall cause to be recorded in the office of the Recorder of__________State of______ ______________, any notice of sale under the Trust Deed/or Mortgage by the_____ statutes of the State of______________relating thereto. Lessee shall furnish to Lessor a complete copy of the Trust Deed and Note secured thereby, together with the name and address of the holder thereof, or a copy of Mortgage if applicable.

D. Lessor agrees that it will not terminate this Lease because of any default or breach hereunder on the part of Lessee if the encumbrancer or the holder or said Trust Deed or Mortgage, within ninety (90) days after service of written notice on the encumbrancer by Lessor of its intention to terminate this Lease for such default or breach, shall:

(a) Cure such default or breach if the same can be cured by the payment or expenditure of money provided to be paid under the terms of this Lease, or if such default or breach is not so curable, and cannot be remedied within said 90 day period, the holder shall diligently pursue to completion steps and proceedings for the foreclosure by sale, or by exercise of a power of sale under and pursuant to the Trust Deed/Mortgage as provided by law; and

(b) Keep and perform all of the covenants and conditions of this Lease requiring the payment or expenditure of money by Lessee until such time as said leasehold shall be sold upon foreclosure, or by exercise of a power of sale, pursuant to the Trust Deed/Mortgage or released or reconveyed thereunder; provided, however, that if the holder of Mortgage or Trust Deed shall fail or refuse to comply with any and all of the conditions of this Article with respect to a breach or default as to which notice of intention to terminate this Lease has been given to the encumbrancer, then and thereupon Lessor shall be released from the covenants of forebearance herein contained with respect to such breach or default.

Any notice to the encumbrancer provided for in this Article may be given concurrently with or after Lessor's notice of default to Lessee as herein provided for in the Article entitled "TERMINATION".

14. TERMINATION: Should Lessee fail to pay any installment of rent or any other sum provided in this Lease to be paid by Lessee at the times herein specified; or should Lessee default in the performance of or breach any other covenant, condition or restriction of this Lease herein provided to be kept or performed by Lessee, and should such default or breach continue uncured for a period of ninety (90) days from and after written notice thereof by Lessor to Lessee, then and in any such event, Lessor may, at its option, terminate this Lease by giving Lessee written notice thereof; subject, however to the rights of any authorized encumbrancer of Lessee as herein set forth in the Article entitled "ENCUMBRANCES", and upon such termination Lessor may, without further notice or demand or legal process, and without prejudice to any other remedy or right of action as herein provided for in the Article entitled "REMEDIES", re-enter and take possession of the leased land and all buildings and other improvements thereon and oust Lessee and all persons claiming under Lessee therefrom, and Lessee and all persons shall quit and surrender possession of the leased land, all buildings and other improvements thereon to Lessor. Lessor expressly waives all other methods and procedures for the termination of this Lease provided by law and agrees that the manner of termination under this Article shall be the sole and exclusive manner of terminating this Lease.

15. RIGHT TO PROCEEDS OF SALE: Upon termination of this Lease in the manner as herein set forth in the Article entitled "TERMINATION", Lessor shall thereafter, pursuant to the provisions of this Article, offer for sale a leasehold estate for a term equivalent to the balance of the unexpired term of this Lease and all improvements on the leased land, subject to the same terms and conditions as set forth in this Lease and subject to the same rights of any encumbrancers as herein set forth in the Article entitled "ENCUMBRANCES". Prior to such sale, Lessor shall give to Lessee twenty (20) days prior written notice of the time and place as herein provided for in the Article entitled "PLACE OF PAYMENTS AND NOTICES". Lessor will post a notice of sale setting forth the time and place of sale on the leased land at least twenty (20) days prior to such sale and will cause a notice setting forth the time and place of such sale to be published at least once, and not less than ten (10) days prior to the sale, in a newspaper of general circulation in_____ _____________County,_______________. Thereafter, the sale of the leasehold estate shall be held in conformance with the notice of time and place of sale and the sale shall be confirmed to the highest bidder. Lessor shall have the right to bid at such sale. Upon any such sale, Lessor shall deduct from the monies derived therefrom the following:

A. The cost of any alteration, repairs, maintenance or redecoration; and

B. All costs of such sale including, without limitation, advertising costs, administrative overhead, commissions and reasonable attorneys' fees incurred; and

C. An amount equal to all delinquent rents, taxes and other charges accruing under the lease to the date of sale with interest thereon and any other legitimate charges against said Leased Land or due Lessor under this Lease.

The then remaining balance of the proceeds from such sale, if any, shall be paid over to Lessee or persons entitled thereto.

16. REMOVAL: Upon the expiration of the term of this Lease, and on condition that Lessee shall not then be in default under any of the covenants and conditions hereof, and not otherwise, Lessee shall have the right during the last ninety (90) days of said term, at its sole expense, to remove from the leased land all buildings and other improvements thereon, and Lessee shall fill all excavations and remove all parts of said buildings remaining after the same are removed and surrender possession of the leased land to Lessor in a clean and orderly condition. In the event any of said buildings and other improvements shall not be removed from the leased land within the time hereinabove provided, the same shall become and thereafter remain a part of the leased land and shall belong to Lessor without the payment of any consideration therefor. Upon the expiration of the term hereof, or any sooner termination of this Lease, Lessee shall execute, acknowledge and deliver to Lessor a proper instrument in writing releasing and quitclaiming to Lessor all right, title and interest of Lessee in and to the leased land and any and all improvements thereon, if not removed by virtue of this Lease or otherwise.

17. PLACE OF PAYMENTS AND NOTICES: All rents and other sums payable by Lessee to Lessor hereunder shall be paid to Lessor at its business office in ______ County, __________. Whenever either party hereto desires to give written notice to the other respecting this Lease, such notice, if not personally delivered to an officer of Lessor, or to Lessee, shall be sent by certified or registered mail, with postage prepaid, and directed to either party at the address hereinabove specified, or at such other address as either party may hereafter designate in writing. The service of any such written notice shall be deemed complete at the time of such personal delivery or within five (5) days after the mailing thereof as herein provided. Should Lessee consist of more than one person, the personal delivery or mailing of such notice to any one of such persons shall constitute complete service upon all such persons. Any notice provided in the Article hereof entitled "ENCUMBRANCES" to be given by Lessor to any encumbrancer of Lessee shall be served in the same manner as herein provided in this Article and shall be delivered to the encumbrancer or directed to its address as last shown on the records of Lessor.

18. REMEDIES: Any termination of this Lease as herein provided shall not relieve Lessee from the payment of any sum or sums that shall then be due and payable to Lessor hereunder or any claim for damages then or theretofore accruing against Lessee hereunder, and any such termination shall not prevent Lessor from enforcing the payment of any such sum or sums or claim for damages by any remedy provided by law, or from recovering damages from Lessee for any default hereunder. All rights, options and remedies of Lessor contained in this Lease shall be construed and held to be cumulative and not exclusive, and Lessor shall have the right to pursue any one or all of such remedies, or any other remedy which may be provided by law, whether or not stated in this Lease. No waiver by Lessor of any breach of any of the covenants or conditions of this Lease by Lessee shall constitute a waiver of any succeeding or preceding breach of the same or any other covenant or condition herein contained. The receipt by Lessor of any rental payment with knowledge of

the breach of any covenant or condition of this Lease shall not be deemed a waiver of such breach, and no waiver by Lessor of any covenant, condition or provision of this Lease shall be deemed to have been made unless expressed in writing and signed by Lessor. Notwithstanding the foregoing provisions of this Article, Lessor's remedies for termination shall be as exclusively provided for in the Article hereof entitled "TERMINATION".

19. REPRESENTATIONS: Lessee covenants and agrees that it has examined the leased land and that the same is delivered to it in good order and condition and that no representations as to said land have been made by Lessor or by any person or agent acting for Lessor, and it is agreed that this document contains the entire agreement between the parties hereto and that there are no verbal agreements, representations, warranties or other understandings affecting the same.

20. HOLDING OVER: This Lease shall terminate and become null and void without further notice upon the expiration of said term. Any holding over shall not constitute a renewal hereof, but the tenancy shall thereafter be on a month to month basis and otherwise on the same terms and conditions as herein set forth.

21. LEASE CREDIT: Lease Credit, as herein provided for in the Articles entitled "EMINENT DOMAIN", "RENTAL ADJUSTMENT", and "OPTION TO PURCHASE", shall be a sum

22. EMINENT DOMAIN:

A. DEFINITION OF TERMS: The term "total taking" as used in this Article means the taking of the entire leased land under the power of eminent domain or the taking of so much of said land as to prevent or substantially impair the use thereof by Lessee for the uses and purposes hereinabove provided.

The term "partial taking" means the taking of a portion only of the leased land which does not constitute a total taking as defined above.

The term "taking" shall include a voluntary conveyance by Lessor to an agency, authority or public utility under threat of a taking under the power of eminent domain in lieu of formal proceedings.

The term "date of taking" shall be the date upon which title to the leased land or portion thereof passes to and vests in the condemnor.

The term "leased land" means the real property belonging to the Lessor, together with any and all improvements placed thereon by the Lessor or to which Lessor has gained title.

B. EFFECT OF TAKING: If during the term hereof there shall be a total taking or partial taking under the power of eminent domain, then the leasehold estate of the Lessee in and to the leased land or the portion thereof taken shall cease and terminate, as of the date of taking of the said land. If this Lease is so terminated in whole or in part, all rentals and other charges payable by Lessee to Lessor hereunder and attributable to the leased land or portion thereof taken shall be paid by Lessee up to the date of taking by the condemnor, and the parties shall thereupon be released from all further liability in relation thereto.

C. ALLOCATION OF AWARD - TOTAL TAKING: All compensation and damages awarded for the total taking of the leased land and Lessee's leasehold interest therein shall be allocated as follows:

(a) The Lessor shall be entitled to an amount equal to the sum of the following:

(i) The fair market value of the leased land as improved (exclusive of the dwelling and appurtenances to such dwelling) as of the date of taking, less a credit to Lessee in the amount as herein set forth in the Article entitled "LEASE CREDIT", discounted by multiplying such remaining fair market value by the factor for the present worth of $1.00 at ____________ percent (_____%) per annum compound interest for the number of years remaining from the date of taking to the next rental adjustment date or to the date of the expiration of the term of this Lease, whichever date is sooner, and

(ii) The present worth of rents due during the period from the date of taking to the next rental adjustment date or to the date of the expiration of the term of this Lease, whichever date is sooner, computed by multiplying the annual rent then payable by the factor for the present worth of $1.00 per annum at ___________ percent (_____%) per annum compound interest (Inwood Coefficient) for the number of years in such period.

(b) The Lessee shall be entitled to the amount remaining of the total award after deducting therefrom the sums to be paid to Lessor as hereinabove provided.

D. ALLOCATION OF AWARD - PARTIAL TAKING: All compensation and damages awarded for the taking of a portion of the leased land shall be allocated and divided as follows:

(a) The Lessor shall be entitled to an amount equal to the sum of the following:

(i) The proportionate reduction of the fair market value of the leased land as improved (exclusive of the dwelling and appurtenances to such dwelling) as of the date of taking, less a credit to Lessee in the amount as herein set forth in the Article entitled "LEASE CREDIT", discounted by multiplying such proportionate reduction in fair market value by the factor for the present worth of $1.00 at ____________ percent (_____%) per annum compound interest for the number of years remaining from the date of taking to the next rental adjustment date or to the date of the expiration of the term of this Lease, whichever date is sooner; and

(ii) The present worth of the amount by which the rent is reduced computed by multiplying the amount by which the annual rent is reduced by the factor for the present worth of $1.00 per annum at ____________ percent (_____%) per annum compound interest (Inwood Coefficient) for the number of years remaining from the date of taking to the next rental adjustment date or to the date of expiration of the term of this Lease, whichever date is sooner.

(b) The Lessee shall be entitled to the amount remaining of the total award after deducting therefrom the sums to be paid to Lessor as hereinabove provided.

E. REDUCTION OF RENT ON PARTIAL TAKING: In the event of a partial taking, the rent payable by Lessee hereunder shall be adjusted from the date of taking to the next rental adjustment date or to the date of the expiration of the term of this Lease, whichever date is sooner. Such rental adjustment will be made by reducing the basic rental payable by Lessee in the ratio that the fair market value of the leased land at the date of taking bears to the fair market value of the leased land immediately thereafter.

23. RENTAL ADJUSTMENT: Upon the expiration of the thirtieth (30th), forty-fifth (45th) and sixtieth (60th) year of the term of this Lease, the rental hereunder shall be adjusted by multiplying the fair market value of the land as improved (exclusive of the dwelling and appurtenances to such dwelling), less a credit to Lessee in the amount as herein set forth in

the Article entitled "LEASE CREDIT", by ________ percent (________%). At the end of said thirtieth (30th), forty-fifth (45th) and sixtieth (60th) year, (as the case may be), and after any such adjustment of rental Lessee shall pay to Lessor such rental as so adjusted during the period applicable thereto at the times and in the manner herein provided for in the Article entitled "RENTAL", provided, however, in no event shall the rental as so adjusted be less than the initial rental in the Article of this Lease entitled "RENTAL". If, upon the expiration of said thirtieth (30th), forty-fifth (45th) and sixtieth (60th) year, (as the case may be), the parties hereto shall have failed to agree upon such adjusted rental, then and thereupon the fair market value of the leased land and the amount of rental to be adjusted in relation thereto as hereinabove provided shall be determined by arbitration as follows:

Within ten (10) days after the expiration of the thirtieth (30th), forty-fifth (45th) and sixtieth (60th) year of said term (as the case may be), each of the parties hereto shall appoint in writing an arbitrator and give written notice thereof to the other party, or in case of the failure of either party so to do, the other party may apply to the Superior Court of _______County, ____________, to appoint an arbitrator to represent the defaulting party in the manner prescribed in the then existing statutes of the State of ___________appliable to arbitration, the provisions of which statutes shall apply to and govern the arbitration herein provided for with the same effect as though incorporated herein. Within ten (10) days after the appointment of said two arbitrators (in either manner) they shall appoint in writing a third arbitrator and give written notice thereof to Lessor and Lessee, and if they shall fail to do so, then either party hereto may make application to said Superior Court to appoint such third arbitrator in the manner prescribed in said arbitration statutes. The three arbitrators so appointed (in either manner) shall promptly fix a convenient time and place in the County of _______for hearing the matter to be arbitrated and shall give reasonable written notice thereof to each of the parties hereto and with reasonable diligence shall hear and determine the matter in accordance with the provisions hereof and of said arbitration statutes, and shall execute and acknowledge their award thereon in writing and cause a copy thereof to be delivered to each of the parties hereto and the award of a majority of said arbitrators shall determine the questions arbitrated, and a judgment may be rendered by said Superior Court confirming said award or the same may be vacated, modified or corrected by said Court at the instance of either of the parties hereto in accordance with said arbitration statutes, and said judgment shall have the force and effect as provided in said statutes. Each of the parties hereto shall pay for the services of its appointee, attorneys and witnesses and one-half of all other proper costs of arbitration. Pending the final decision of such adjusted rental, Lessee shall pay to Lessor the amount of rent previously payable under the Article of this Lease entitled "RENTAL". If such adjusted rental as finally determined shall exceed the amount of the previous rental, the excess amount accruing during the interim period shall be paid by Lessee to Lessor within thirty (30) days after the final determination of said adjusted rental. If such adjusted rental as finally determined shall be less than such previous rental the amount of any excess paid by Lessee during said interim period shall be credited against the first rentals thereafter payable hereunder.

24. DRAINAGE AND FILL: Lessee shall cause all drainage of water from the leased land and improvements thereon to drain or flow into adjacent streets and not upon adjoining property, and Lessee shall so maintain all slopes or terraces on the leased land as to prevent any erosion thereof upon such streets or adjoining property. Lessee will make its own tests to ascertain the amount and extent of the present fill and any subsurface or soil condition upon or in connection with the leased land, and this Lease is made subject to and without any liability on the part of Lessor for any damage resulting from any fill or any subsurface or soil condition upon or in connection with the leased land or adjacent property.

25. ANTENNAE: Lessee shall not erect and maintain or permit to be erected or maintained upon the leased land any tower, antenna, aerial or other facility for the reception or transmission of radio or television broadcasts or other means of communication except by installation inside of the dwelling house constructed on said premises or by underground conduits.

26. COMMUNITY ASSOCIATION:

Prior to the execution of this Lease there has been organized under the laws of the State of ___________ a non-profit corporation known as _______ ____, ("Association" herein) for the purpose of maintaining community facilities, architectural control and other community services on behalf of its members. As a material part of the consideration of this Lease and as an express condition to the continuance of any of the rights of Lessee hereunder, Lessee covenants and agrees to become and during the term of this lease to remain a member in good standing of said Association, and to abide by the Articles of Incorporation, Bylaws and rules and regulations of said Association, now or hereafter existing, and to pay to said Association before delinquency all dues, fees, assessments and other charges from time to time duly and regularly levied or assessed by said Association in furtherance of its community purposes. Any default by Lessee in the performance of any of the foregoing covenants and conditions shall constitute a material failure of consideration hereunder and a breach of the covenants and conditions of this Lease.

Provided, however, that the foregoing provisions of this Article shall not apply to any established lending corporation which as authorized encumbrancer acquires title to the leasehold under this Lease by or in lieu of foreclosure for a period of three (3) years from the date of so acquiring title, provided that such encumbrancer shall (a) apply in good faith to such Association for membership and shall be denied or refused membership, or once having become a member such membership shall be revoked for any reason other than the failure to pay dues, fees or assessments or other charges duly and regularly levied or assessed by said Association; and (b) pay all dues, fees, assessments and other charges duly and regularly levied or assessed by said Association against the members thereof, whether or not said encumbrancer becomes a member of said Association. The provisions of the first paragraph of this Article shall be binding upon and shall be performed by any transferee or assignee of said encumbrancer.

27. USE OF COMMUNITY FACILITIES: During the term of this Lease and so long as Lessee shall not be in default under any of the covenants, conditions or restrictions of this Lease, and not otherwise, Lessee and members of Lessee's family and Lessee's guests, invitees and authorized subtenants, shall have the privilege, jointly with members and other authorized persons, to use said parks and other community facilities, subject to the right of said Association therein and to the Bylaws and rules and regulations of said Association from time to time adopted by it; provided, however, that Lessor shall not be liable for any injury or damage to Lessee or any of said other persons caused by or arising from the use of any of said community facilities as herein provided for in the Article entitled "LIABILITIES", each and all of the provisions of which Article shall apply with equal force and effect to each and all of said community facilities.

28. ASSUMPTION OF COMMUNITY SERVICES: If at any time during the term of this Lease said Association shall fail or cease to maintain said community facilities, the Lessor may, at its option, assume the performance of the community services of said Association. In such event, Lessee agrees to pay to Lessor, in addition to the rental as herein provided for in the Article entitled "RENTAL", Lessee's pro rata share of Lessor's costs of the maintenance and operation of said community facilities and services including the fee of a management agent or a reasonable fee charged by Lessor for its services in management and operation of the community facilities. Said costs shall be

determined annually from the financial records of Lessor, which records shall be conclusive for the purposes of this Article, and lessee's pro rata share shall be determined by dividing such costs equally by the number of residential lots within all tracts, the occupants of which shall be entitled to the use of said community facilities. Lessee's pro rata share of said annual costs determined as aforesaid shall be paid by Lessee to Lessor within thirty (30) days after written notice thereof by Lessor to Lessee, and any default in the payment thereof shall constitute a breach of the covenants and conditions of this lease.

29. ENCROACHMENTS: If a dwelling house is constructed on the leased land, the wall or walls of which adjoin the wall or walls of a dwelling constructed on a contiguous lot, any such wall shall be considered to adjoin and abut the wall of the contiguous lot against the surface from the bottom of the foundation over the full length and height of any building so erected for residential purposes. Both Lessee and the lessees of contiguous lots shall have a reciprocal easement appurtenant to each of said lots over said contiguous lots for the purpose of accommodating any encroachment of any wall of any dwelling house.

Lessee and the lessees of contiguous lots shall have a reciprocal easement appurtenant to each of said lots over said contiguous lots for the purpose of accommodating any natural settlement of any structures located on any of said lots.

Should there be found to exist any party wall or party fence, the agreement between Lessee and the lessee of a contiguous lot or lots shall be that the lessees of the contiguous lots who have a party wall or party fence shall equally have the right to the use of such wall or fence, and such wall shall be considered to adjoin and abut against the surface from the bottom of the foundation over the full length and height of any building so erected. Such rights of use shall be as not to interfere with the use and enjoyment of the lessees of adjoining lots, and in the event that any such party wall or fence is damaged or injured from any cause other than the act or negligence of one of the lessees, the same shall be repaired or rebuilt at their joint expense.

30. CONSTRUCTION AND EFFECT: Time is of the essence of this Lease. The article headings herein are used only for the purpose of convenience and shall not be deemed to limit the subject of the articles hereof or to be considered in the construction thereof. Each and all of the obligations, covenants, conditions and restrictions of this Lease shall be deemed as running with the land and shall inure to the benefit of and be binding upon and enforceable against, as the case may require, the successors and assigns of Lessor, and subject to the restrictions of the Article hereof entitled "ASSIGNMENT", the heirs, executors, legal representatives, encumbrancers, assignees, successors and subtenants of Lessee. If Lessee consists of more than one person, the covenants and obligations of Lessee hereunder shall be the joint and several covenants and obligations of such persons. In this Lease, the masculine gender includes the feminine and the neuter, and the singular number includes the plural, whenever the context so requires.

31. LEASE SUPERSEDED: The Lease dated ______________________________ covering the premises described in Paragraph 1 hereof which was recorded on ____________________________________in Book ___________, Page ___________, of Official Records of_______County,___________, is hereby cancelled and superseded by this Lease.

32. OPTION TO PURCHASE: Lessor hereby grants to Lessee an option to purchase the leased premises at any time during the term of this Lease if Lessee is not in default hereof on the following terms and conditions:

A. The purchase price shall be the fair market value of the land as improved (exclusive of the dwelling and appurtenances thereto and without consideration of the fact that the land is leased), less a credit to Lessee in the amount as herein set forth in the Article entitled "LEASE CREDIT"; provided, however, in no event shall the purchase price be less than the annual rental then payable under the Lease divided by __________ percent (__________%). In the event Lessor and Lessee are unable to agree, the fair market value of the land shall be determined by arbitration as herein provided for in the Article entitled "RENTAL ADJUSTMENT".

B. The land shall be conveyed by Grant Deed containing the provisions set forth in the Corporation Grant Deed marked Exhibit "A" attached hereto and made a part hereof.

C. The purchase shall be consummated through a thirty (30) day escrow to be opened promptly after Lessee's giving written notice to Lessor of its exercise of the option, through which escrow Lessor will deliver a standard abstract or a Land Title Association owner's form of policy of title insurance covering the land, and a deed thereto to Lessee, and through which escrow Lessee shall deliver the purchase price in cash to Lessor. The escrow shall be opened at an escrow company or law firm, et al, and the Lessor shall pay for the title policy, one-half of the escrow fee, and normal seller's expenses, and Lessee shall pay for one-half of the escrow fee and purchaser's normal expenses. The escrow instructions shall be on a standard form of such escrow company.

D. The land shall be conveyed subject to the approval of any lender having an encumbrance of record on the leasehold estate.

IN WITNESS WHEREOF, the parties have executed this Lease With Option to Purchase, the Lessor by its officers thereunto duly authorized and its corporate seal affixed, as of the day and year first above written.

THE LANDGUIDE TRUST CORPORATION,
A State Chartered Corporation
under Trust No. 14S-7211-02

By: LANDGUIDE ENTERPRISES
as Attorney-in-Fact

By______________________	By______________________ Vice President
By______________________	By______________________ Assistant Secretary
"LESSEE"	"LESSOR"

STATE OF__________________)
__________________________) SS.
COUNTY OF_________________)

On ________________________, before me, the undersigned, a Notary Public in and for said State, personally appeared ______________________________ ______________________________ known to me to be the person(s) whose names is/are subscribed to the within instrument and acknowledged to me that _he_ executed the same.

WITNESS my hand and official seal.

____
Notary Public in and for said State

STATE OF_______________)
_______________________) SS.
COUNTY OF______________)

On________________________________, before me, the undersigned, a Notary Public in and for said county and state, personally appeared______________________ ____________________and____________________________, known to me to be the Vice President and Assistant Secretary, respectively, of Landguide Enterprises, a corporation, and known to me to be the persons who executed the within instrument on behalf of the corporation, whose name is subscribed to the within instrument as the attorney in fact of The Landguide Trust Company, a corporation, as Trustee under Trust No. 14S-7211-02, and acknowledged to me that it subscribed the name of The Landguide Trust Company, a corporation, as Trustee under Trust No. 14S-7211-02 thereto as principal and its name thereto as attorney in fact.

WITNESS my hand and official seal.

Signature____
Notary Public for said county and state.

WHEN RECORDED, MAIL TO:

Space above this line for Recorder's use

LESSORS' CONSENT TO ASSIGNMENT OF SUBLEASE BY TRUST DEED/MORTGAGE

The undersigned, Lessor by mesne assignments and amendments of record in the Sublease covering the following described real property situated in the County of________, State of ________, to-wit: Lot____, Tract_____(hereinafter referred to as "said Lease") does hereby consent to the assignment by the Lessee under said Lease of the leasehold estate thereunder by Deed of Trust dated __________________(hereinafter referred to as "TD/Mortgage") in favor of ____
____
____
____
____
____(hereinafter called "Encumbrancer") to secure a note in the principal sum of $__________and other obligations set forth in the Trust Deed/Mortgage and which is recorded concurrently herewith in the office of the County Recorder of __________County,________________. The foregoing consent is made and accepted upon and subject to the following covenants and conditions, each and all of which shall be binding upon and inure to the benefit of the encumbrancer and its transferee(s), to-wit:

(a) Said Trust Deed or Mortgage as acquired thereunder shall be subject to each and all of the covenants, conditions and restrictions set forth in said Lease and to all rights and interests of the Lessors, thereunder, except as herein otherwise provided;

(b) In the event of any conflict between the provisions of said Lease and the provisions of the Trust Deed or Mortgage, then said Lease shall control as follows:

(c) The prior written consent of Lessor shall not be required:

(1) To a transfer of said Lease at foreclosure sale under the trust deed, or a mortgage foreclosure, or by an assignment in lieu of foreclosure; or

(2) To any subsequent transfer by the encumbrancer if the encumbrancer is an established bank, savings and loan association, or insurance company, and is the purchaser at each foreclosure sale;

Provided, that in either such event the encumbrancer forthwith gives notice to the Lessor in writing, of any such transfer, setting forth the name and address of the transferee, the effective date of such transfer, and the express agreement of the transferee assuming and agreeing to perform all of the obligations of said Lease, together with a copy of the document by which such transfer was made; and the payment to Lessor of the transfer fee provided in said Lease, but in any event not more than $__________.

Any transferee under the provisions of subparagraph (c) above shall be liable to perform the obligations of the Lessee under said Lease only so long as such transferee holds title to the leasehold. Any subsequent transfer of the leasehold shall not be made without the prior written consent of the Lessor and shall be subject to the conditions relating thereto as set forth in said Lease.

CAMEO CONSENT TO ASSIGNMENT OF
SUBLEASE BY TRUST DEED
46/10-26-77

(d) Upon and immediately after the recording of the TD/Mortgage, Lessee, at Lessee's expense, shall cause to be recorded in the office of the Recorder of ______________ County, ______________, a duly executed and acknowledge written request for a copy of any Notice of Default and any Notice of Sale under the TD/Mortgage, as provided by the statutes of the State of ______________ related thereto. Concurrently with the execution of said consent, Lessee shall furnish to Lessor a complete copy of the TD/Mortgage and Note secured thereby, together with the name and address of the holder thereof.

Lessor agrees that it will not terminate said Lease because of any default or breach thereunder on the part of Lessee, if the encumbrancer or the trustee under such TD/Mortgage, within ninety (90) days after service of written notice on the encrumbrancer by Lessor of its intention to terminate said Lease for such default or breach, shall

(1) Cure such default or breach if the same can be cured by the payment or expenditure of money provided to be paid under the terms of said Lease, or, if such default or breach is not so curable, cause the encumbrancer on a TD/Mortgage to commence and thereafter to diligently pursue to completion steps and proceeedings for the exercise of the power of sale under and pursuant to the TD/Mortgage in a manner provided by law;

(2) Keep and perform all of the covenants and conditions of said Lease, requiring the payment or expenditure of money by Lessee until such time as said leasehold shall be sold upon foreclosure pursuant to the TD/Mortgage or shall be released or reconveyed thereunder, or shall be transferred upon judicial foreclosure or by an assignment in lieu of foreclosure;

provided, however, that if the holder of a TD/Mortgage shall fail or refuse to comply with any and all of the conditions of this paragraph, then and thereupon Lessor shall be released from the covenants of forebearance herein contained.

Any notice to the encumbrancer provided for in this paragraph may be given concurrently with or after Lessor's notice of default to Lessee as provided in said Lease.

In the foregoing consent the masculine gender includes the feminine and neuter, and the singular number includes the plural, whenever the context so requires.

DATED: ______________________________

The Landguide Trust Corporation a
state chartered corporation, as Trustee
under Trust No. 14S-7211-02

By: Landguide Enterprises
As Attorney-in-Fact

By ______________________________
Carrie A. Stampe, Assistant Secretary

State of ______________) ss
Couny of ______________)

On ______________, before me, the undersigned, a Notary Public in and for said county and state, personally appeared ______________, known to me to be the Assistant Secretary of Landguide Enterprises, a corporation, and known to me to be the person who executed the within instrument on behalf of the corporation, whose name is subscribed to the within instrument as the attorney-in-fact of the Landguide Trust Corporation, a corporation, as Trustee under Trust No. ______________, and acknowledged to me that it subscribed the name of The Landguide Trust Corporation, a corporation, as Trustee under Trust No. ______________ thereto as principal and its name thereto as attorney-in-fact.

WITNESS my hand and official seal.

Signature ______________________________
Notary Public for said county and state

WHEN RECORDED MAIL TO:

____

(Space above this line for Recorder's use)

ASSIGNMENT

For valuable consideration, receipt of which is hereby acknowledge, the undersigned __________
____
____
does hereby transfer and assign to ______________________________
____
____
all right, title and interest of the undersigned, as Lessee, in and under that certain Lease dated ______
by and between LANDGUIDE ENTERPRISES, a state corporation, as Lessor and __________
____
as Lessee, recorded on ____________ ______________ in Book ______________________
Page of Official Records of County, ____________ described as follows, to wit:

TOGETHER WITH all buildings and other improvements on said land.

DATED: ______________________________
____
(Assignor)

STATE OF__________________) ss
COUNTY OF_________________)

On ______________________________,
before me, the undersigned, a Notary Public in and for said County and State, personally appeared ______________________________
____
____
known to me to be the person___whose name_ is/are subscribed to the within instrument and acknowledged to me that ___ he ___ executed the same.

WITNESS my hand and official seal.

____
Notary Public in and for said County and State

STATE OF__________________) ss
COUNTY OF_________________)

On ______________________________,
before me, the undersigned, a Notary Public in and for said County and State, personally appeared ______________________________
____
____
known to me to be the ______________________
President, and ______________________________
____
known to me to be the ______________________
Secretary of the Corporation that executed the within Instrument known to me to be the persons who executed the within Instrument on behalf of the corporation therein named, and acknowledged to me that such corporation executed the within Instrument pursuant to its By-Laws or a resolution of its Board of Directors.

WITNESS my hand and official seal.

____
Notary Public in and for said County and State

MAIL TAX STATEMENTS TO: ____
(Name) (Address)

Assignment—Page 1 of 2
47/10-26-77

CONSENT AND CONFIRMATION

The undersigned, as lessor under the lease referred to in the foregoing assignment, hereby consents to such assignment without, however, waiving the restrictions, if any, contained in said lease with respect to future assignments thereunder, and confirms that the provisions of said lease with respect to assignment thereof have been complied with, and hereby releases the assignor as lessee of said lease from any and all further obligations thereunder and hereby accepts the assignee as lessee under said lease, to all intents and purposes as though said assignee was the original lessee thereunder.

DATED:

The Landguide Trust Corporation, a
state chartered corporation, as Trustee
under Trust No. 14S-7211-02

By: Landguide Enterprises
As Attorney-in-Fact

By____________________________
Assistant Secretary

STATE OF______________________) ss.
COUNTY OF____________________)

On ______________________________, before me, the undersigned, a Notary Public in and for said county and state, personally appeared ____________________________
____________________________________, known to me to be the Assistant Secretary of Landguide Enterprises, a corporation, and known to me to be the person who executed the within instrument on behalf of the corporation, whose name is subscribed to the within instrument as the attorney-in-fact of the Landguide Trust Corporation, designated as Trustee under Trust No. 14S-7211-02, and acknowledge to me that it subscribed the name of the Landguide Trust Corporation, designating it as a Trustee under Trust No. 14S-7211-02 thereto as principal and its name thereto as attorney-in-fact.

WITNESS my hand and official seal.

Notary Public in and for said County and State

Assignment—Page 2 Rev. 10/77 47/10-26-77

ACCEPTANCE AND AGREEMENT

The undersigned assignee named in the foregoing assignment (if more than one, then jointly and severally) hereby accepts said assignment and hereby agrees with and for the benefit of the lessor, under the lease described in said assignment, to keep, perform and be bound by all of the terms, covenants and conditions contained in said lease on the part of the lessee therein to be kept and performed, to all intents and purposes as though the undersigned assignee was the original lessee thereunder.

Dated ______________________________

Address: ______________________________

(Assignee)

STATE OF____________________) ss
COUNTY OF__________________)

On ______________________________,
before me, the undersigned, a Notary Public in and for said County and State, personally appeared______________________________
____
____
known to me to be the person___whose name___ is/are subscribed to the within instrument and acknowledged to me that___he___executed the same.

WITNESS my hand and official seal.

Notary Public in and for said County and State

STATE OF____________________) ss
COUNTY OF__________________)

On ______________________________,
before me, the undersigned, a Notary Public in and for said County and State, personally appeared______________________________
____
____
known to me to be the______________________________
President, and______________________________
____
known to me to be the______________________________
Secretary of the Corporation that executed the within Instrument known to me to be the persons who executed the within Instrument on behalf of the corporation therein named, and acknowledged to me that such corporation executed the within Instrument pursuant to its By-Laws or a resolution of its Board of Directors.

WITNESS my hand and official seal.

Notary Public in and for said County and State

11

Financing The Special Purpose Project

As a builder, speculator, land developer or member of an investment group, you must be able to confront your potential lender with essential information to go along with your building plans. Your ability to familiarize yourself with special purpose property goes hand in hand with knowing how to obtain loans for this property. This chapter outlines the procedure to obtain building funds for such overlooked special purpose property as churches and motels. This procedure can be applied to other types of special purpose property common to mortgage lending.

Property is classified economically as to its highest and best use. The ability to visualize this highest and best use for any given property, and then to acquire it at less than its worth, is the mark of the successful operator. Property for building purposes can achieve its highest and best use in the following categories.

Residential
Commercial
Industrial
Recreational
Educational
Political
Religious

The most important factor to consider in any of the above categories is the intensity of use. It may be that land in that category is underimproved, overimproved or sufficiently improved to its highest and best use. Naturally you must use more care in analyzing property that is expensive and scarce.

How To Obtain A Special Purpose Property Loan

Most special purpose lending is made by commercial banks, insurance companies, syndicates and colleges that sometimes deal in participating mortgages.

Lenders make construction loans for special purpose property primarily to develop a permanent mortgage investment. Special purpose loans are usually made to operating builders involved in the construction of very large projects. For construction loan type mortgages in which only "building" money is required, the building loan agreement is made between the borrower and the lender well before construction begins. Provided the builder proceeds with construction in accordance with the agreed plans and specs and all prior liens and encumbrances are cleared, the lender advances money at various stages of progress. The loan agreement usually carries an affidavit executed by the builder detailing specific money which will be deducted from proceeds of the loan for title examination, taxes, assessments, recording fees, pay-off of existing mortgages or trust deeds and the like. The net amount is then shown as the available amount for the project. Many states require that the building loan agreement be recorded well in advance of the actual first payment on the loan.

When the owner of land being sold to the builder agrees to a certain amount of cash to be paid to him, the builder must produce credit for the balance of the price as well as the construction money for the project. So the builder contracts to purchase the land from landowner by giving him the balance of the sales price in the form of a purchase money mortgage. Upon payment of stipulated amounts, portions of the mortgaged land can be released according to the contract. Now that the land has been provided for the project, the builder goes to the financial institution and makes application for a construction mortgage loan. When the lender's feasibility study is complete showing the site's highest and best use as he sees it, and when an analysis of the plans and specs is finished, the appraiser and loan officer will finish the details before the construction loan commitment is made.

The initial portion of the mortgage money is usually the largest portion of the loan. The reason for this is because it is advanced on the security of the land plus the building as in an advanced stage of the proposed construction. Once this advance is made, the landowner's purchase money mortgage is paid off and the land is released for groundbreaking.

The Church Mortgage Loan

It is rare for a builder, developer or subdivider to seek land to develop special purpose property to its highest and best use for the resale market. Since the church is a social factor in family life, and the use of the church has been accepted as a community social center, tract developers often reserve land for the exclusive use of special religious structures. Depending on the subdivision lender at the time of the original construction, little interest may be shown by lending institutions when it comes time to build the church buildings. Vacant land could end up being sold at a reasonable price to a group of homeowners who are ready to form their congregation. Since a presence of adequate religious facilities is thought by appraisers to contribute to social stability, a new church construction tends to increase property values. The need to borrow for church construction would not appear to be a problem. Unfortunately, it is very often a specific problem; as noted before, unless the original subdivision mortgage lender-investor is especially enterprising, an outside special purpose construction loan may be necessary.

Legal problems also cause lenders to hesitate in making a

mortgage loan secured by churches. The lender's attorney is required to become acquainted with the legal structure of the congregation's specific denomination. In some cases the denomination's title to the property is held by the local parishes while in other denominations it is owned by the district or central conference. Sometimes the local parish must obtain the approval of other bodies to borrow the money, even though that parish actually owns the property. In cases of this kind, lenders find they must gain the approval for a loan at a general meeting of the entire congregation which requires the lenders' presence.

Some religious denominations are highly organized and borrow through large bond issues. Bond specialists sell such church bonds to insurance companies, pension funds, banks and institutional investors. The operation of school systems, homes for the aged, orphanages and hospitals makes larger denominations very good customers of the bond specialists. Smaller denominations have church extension boards and home mission boards which help newly established congregations with grants and loans, but the churches themselves must usually obtain first mortgage loans for amounts above those granted by these church groups.

Loan brokers are valuable in this type of loan. A mortgage loan broker knows that the LVR is quite low—seldom more than 60% of the construction cost, noting the land is already owned by the church. The balance, or 40% of the cost in cash or equity, must be vested in the congregation. The broker's sources of funds may include banks, savings and loans, mortgage companies, life insurance companies or private investors. He tries to secure the loan at one of these places, but he usually knows in advance where his chances of success are best.

Church Loan Requirements

It pays builders to be aware of this type of loan problem. If you know the requirements for church loans you may be able to find your own lender and win a contract in this specialty area. The usual church construction loan made by nearly all institutional lenders looks like this:

1. *Loan amount:* About 50% loan to appraised value, and in some cases 60%.

2. *Loan term:* Usually about 10 years maturity.

3. *Interest rate:* Controlled by supply and demand of available funds.

4. *Fees:* Comparable to current construction loan fees of other structures but an appraisal fee is usually high depending on the amount of architectural and construction work to be done.

5. *Construction advances:* Certificates from the supervising architect on all advances are an important requirement. The lender's inspector may make weekly inspections of the progress. Labor and material advances are in one lump sum disbursement of up to 90% of the installed portion.

6. *Escrow payment:* Since religious structures are tax exempt, there are no payments collected other than the full paid insurance policy or policies for the full insurable amount. These policies must be for a minimum of one year. The first payment on the loan, payable directly to the lender, may be due as far ahead as one year from the date of escrow closing or the date of the note. Interest is figured from the date of the advance to the date the first payment is due on the note.

7. *Architect's supervision:* The architect's supervision of the job should be closely controlled. A certificate is required actually noting a stage of construction, the approval of disbursements, and the completion certificate in advance of the last payment.

8. *Completion bond:* The congregation's construction contract must be covered in full by a completion bond with the lender's (mortgagee's) name incorporated in the language of the bond and giving the lender the same right to act as the borrower (mortgagor) but with no more rights than the congregation.

9. *The budget report:* At least two years' financial history should be represented in the church budget. A projected budget for the period following the loan should be available. The proposed budget must include the payment on the loan as well as the anticipated operating cost. A lender may ask for the list of pledges to the building fund, if any, payable for two or three years hence. Smaller church loans require only a limited liability guaranty of the debt by a few responsible members of the congregation.

10. *Community loan:* Some risk rate criterion or formula for the value of a church loan is usually established by lenders just as value measures are established in other types of loans. The physical location of the church, its value, the estimated value of the homes in the areas adjacent to the church, an estimate of the economic picture for the area, the type of membership, are all included in the formula. The formula is usually created for a loan value per 50 families. Formulas of the past have been as low as $100,000 in loan value per 50 families, assuming three people per family. So $100,000 per 50 families might be a guide to church loans. Changing circumstances may make this only a preliminary guide for religious lending.

The Motel Mortgage Plan

We can trace a special purpose loan from its inception as a developer's idea to a property valuation with which you can confront a lender. Included below are the builder's steps to finding value through three types of valuation—the market approach, the cost approach. and the income approach.

You have made contact with some businessmen who plan to build a 100 unit motel with restaurant. A lender is given the details of the development, and you anticipate an approximate figure for construction this lender can make available. The general "commitment" from the lender turns out to be incompatible with the developers' plans; further conferences with the parties involved, including the staff appraiser, reveal how little was actually known about the plans when they were submitted to the lender. It is agreed that a feasibility report must be done. Usually the mortgagee recommends an independent appraiser for feasibility reports for larger projects. In this case the report substantially raises your rule of thumb "commitment" from the lender's original LVR. Basic, dependable fair market value is established for a realistic construction loan for the motel project. Not all projects should have feasibility reports; but you must supply a lender with more information than merely a good idea of the "highest and best" use of the land.

As with any other large development, you must classify your motel plan before your visit to the lender. Two broad classifications of motels are the transient and the terminal. Transient facilities are patronized by the traveler enroute, while those who plan to stay for longer periods at a resort or in a commercial area select a residential hotel or motel. The promoter must have a knowledge of the area. The economic life of the enterprise depends upon your knowledge of the direction of growth and stability of the community. Since motels are a type of commercial property, capitalizing net income this year is not necessarily next year's valuation.

The lender should be expected to know exactly what competition exists in your areas and how your project would affect the motel occupancy factor for the area. He should be expected to come up with such an analysis. But it is a simple matter for you to determine how many total motel units are available in the community. Learn how many of these units are modern and what facilities or services are offered by the competition. At this point you may learn whether or not another motel in the area would gain a high enough occupancy rate to justify it economically. How many have pools? Most important, how many have restaurants as part of the motel accommodations, since your friends are planning to include this facility? Your local real estate board or chamber of commerce is a source of information. The number one consideration for a motel is location. No doubt your potential clients have covered some of these matters. How sensitive are they to this important factor? A few short blocks can mean the difference between success and failure. When it comes to discussing the amount of a loan for construction, good points such as location, facilities and projected high occupancy factors are selling points. Sometimes a builder views bad points indifferently as far as his contract is concerned—just as long as he gets the job is his attitude. But by making sure you cover all the points with your investors you assure yourself of a good chance at the job. Otherwise, you can only guess why the contract fell to someone else.

The right location for a transient motel restaurant could be wrong for the terminal one. How far away are quality restaurants that could accommodate for sporting events or conventions nearby? Convenience and access to the site can be visualized before construction merely by noting the traffic signal lights and the divided highway, if any. Look for potential blockage of the entrance to the property, or the intensity of the traffic nearby. "Special purpose" has a special meaning in the case of the motel loan, in that the building may or may not be the type that can be converted. The building is more valuable from a lender's viewpoint if the property is one that, with inexpensive modifications, may be used in other businesses.

Although the builder-contractor is not particularly affected by the future business to be conducted within the structure, he should be vitally concerned with the generosity of the loan-to-value ratio and should look for a competitive interest rate. The lender will not only investigate the background characteristics and collateral capacities of your contracting associates, but he will ascertain whether or not the applicant is indeed a good businessman and has the ability to make money. Will he personally manage the business or will he delegate the management to others?

This is true of any applicant interested in commercial building construction, whether it is a motel, a shopping center, or an industrial park. The lender must know whether the special purpose business property is being built for lease. Insurance companies have been very active in these commercial loans as well as in industrial property built for lease. Many financially sound businesses lease space in popular shopping centers that house from four to 12 shops. Local lenders are usually very interested in this type of commercial loan for construction as leases are considered basic value. Some savings associations make loans on the smaller shopping centers in smaller communities. Little interest is shown by the insurance companies for the suburban shopping center, however.

We have used the motel as an example of special purpose commercial development property. Like any business, it has income. Architecturally and functionally, it has a special purpose. In principle, a motel valuation is the same as any other type of income producing property valuation. The applicant's character, business capacity, and collateral are basic considerations to the lender in his evaluation of any real estate loan, no matter what kind of income-producing property is used as security for the construction loan or refinance.

A motel income is a simple problem in valuation. You must arrive at a salable, realistic valuation that can be used as collateral in the construction loan. This collateral capacity is what you are determining in the following value calculations. The steps below can be applied with little variation to almost any capitalization approach to business construction loans. You can use them to arrive at the potential value of any proposed commercial project when operating with other people's money. This valuation must be determined before you enter the lender's office and involves three approaches.

So far, you have earmarked the vacant land suitable for your project, which is either owned or being acquired by your client. You are seeking value information for the best construction loan commitment you can find for a client—thus arming yourself for the building contract.

Step One - Market Approach

For the market approach, you must find comparable sales. This is not easy, nor is it the complete answer; but you are lost without some comparable estimate of value in your area. A full feasibility report by a professional appraiser involves comparables. But all three of these approaches cover preliminaries prior to seeing the lender, and are not an appraisal. Make a complete check for sales of modern motels in your locality. In this example, you are able to find only two recent motel sales close enough to your project to represent the market for your area.

Sale A is an 80 unit motel with a swimming pool and coffee shop. It consists of near-modern construction with average maintenance. Dependable information indicates that the furnished property sold for $1,020,000. This breaks down to $12,750 per unit. The motel is seven years old at the time of the sale. It has a poor location compared to your proposed 100 unit site.

Sale B is a one-year-old 68 unit motel with swimming pool. It has a modern design, equal or slightly superior construction to your design and a fine restaurant facility. The restaurant was not included in the price. The location is equal in status to your proposed site. You have confirmed this sale at $1,070,000, or $15,735 per unit. Further breakdown reveals that this price reflects an income multiplier of 8.21 x annual net income after deducting all costs. Adjustments are made for the differences in date of sale, terms, location, size of site, age and condition of the improvements. Your conclusion is that B is the most comparable property with its quality construction and the nearby facilities of a superior restaurant.

By taking the market approach, you conclude that your proposed motel project should be minimized at $15,735 per unit or 100 units at $15,735 per unit totals $1,573,500.

Step Two - Cost Approach

Ideally, your motel construction cost should be compared to a recent motel construction. Construction and labor cost tables can be applied to your motel proposal, especially if comparable motels have been constructed within the last few months nearby. But you can only compare if you assume that a total construction cost (not sale cost) is obtained. This would be a total square foot cost for the motel rooms. Do this by dividing the overall construction costs of a recently built motel by the number of rooms, thereby arriving at a per unit cost. Say your available cost tables show a per unit cost of $23.00 per square foot, and a total cost for your own project is found to be $1,290,300. If the known total costs of recently built comparables are accurate, your square foot costs can be adjusted for the comparable's dissimilarity of unit cost.

Costs to build do not necessarily equal value if poor design, cheap construction, misplaced improvements or inadequate parking is allowed. Your lender will be concerned with these factors in addition to cost factors and the value of the land. If your client just bought the land, the (lineal) front foot cost is known. If not, check the most recent land sales in the area. Say you have found sales indicating a price range from $1,300 per front foot up to $1,450. You check each sale for the same highest and best use as well as for size and location. After checking the sales you settle on a project site land value of $1,500, or 168 feet at $1,500 per front foot equalling $252,000.

100 units at 561 square feet per unit	$1,290,300
168 feet of improved front footage	252,000
Total	$1,542,300

This cost approach does not include units as furnished for rental occupancy, nor for the restaurant structure. It includes contractor's overhead and profit as viewed by the mortgagee (lender) at $23 per square foot. You can conclude that your proposed motel project should justify a value estimate for land and building (units) of $1,542,300 by the construction *cost approach*, or $15,423 per unit without furnishings.

Step Three - Income Approach

Although market and cost estimates are important conversational figures for your lender's preliminary information, he will be thinking also of the earnings your proposed motel project may realize. So that the lender may capitalize at a current local rate, a short income projection may be done before his trip to the proposed site. This will produce an estimate of value by the income approach.

If you have no in-house data available as to income and expenses of your proposed business, you must again depend on outside data obtained from comparables. This means that managers or accountants of motel B in your market approach must cooperate with your efforts. This data for 68 units must be converted to a 100 unit basis for your proposed site.

Gross Income 100 Units Annually	
@$17.80 average daily rate or	
$1,780 X 365 equals	$650,000
Less vacancy 21.2%	137,800
Anticipated income	$512,200
Anticipated expenses	321,762 *
Annual net income	$190,438

Operating Expenses	
Administrative	$240,480
Building maintenance	11,516
Furniture and equipment	9,040
Taxes and insurance	31,720
Wear and tear reserves	29,006
	$321,762 *

Using Sale B's multiplier of 8.21 times the net income based on B's sale price in the open market, you can now convert to a dependable capitalization rate for the 100 units as proposed. The multiplier is the reciprocal of the rate, or one (unit) divided by the multiplier: (1 ÷ 8.21) = .12 or 12%. Refer to market approach, Sale B. Capitalizing the above net income results in $1,586,983.

$$\$190,438 \div .12 = \$1,586,983$$

The loan officer appreciates your developing a few rules of thumb before coming in to discuss the loan specifics. He is not a graduate engineer himself, but he does not have to be a registered architect to know the costs per square foot for commercial, warehouse and other types of structures in addition to residential property. Your figures give you the opportunity to prove the merits of the particular loan for you and your client. What you are after is your own value opinion of your proposed project to land the building contract.

Broker's Net

Making your own "feasibility" report should reflect to the lender your confidence in the success of the project and in the success of the loan application when you reach a realistic valuation figure. Competition in the area of commercial or industrial mortgages is intense, since insurance companies are very much in the market for loans on this type of property. In our example, the cost approach has been purposely simplified; the added cost of new furniture or other motel equipment would bring the final value closer to that found by the income and the market approaches. A lender thinks in terms of land and building.

Here are the three value figures you can present to the lender along with the good points of location, access, and the like.

Market approach	$1,573,500
Income approach	$1,586,983
Cost approach	$1,542,300 (unfurnished)

Any complete and detailed appraisal will not match hypothetical figures, but if you follow the pattern demonstrated in all three approaches you are at least "in the ballpark." Sometimes the lender likes to use a "broker's net" as he talks with an applicant for a special purpose business mortgage loan. In this case, his rule of thumb may merely include gross income of motel B in your survey less taxes, insurance and vacancy factor. This is say, $330,477 per year, broker's net. His calculator should show a "broker's capitalization rate" of 30.8% (or 330,477 ÷ 1,070,000). If studies of other similar recent sales indicate approximately the same rate he may use this rate in his determination of your proposed project's worth before he decides to accept the application. This will be approximately $1,560,000, or $480,480 ÷ .308.

12

Condominiums And Cooperative Apartments

The details of special purpose property lending would cover a career of financial research. But we can consolidate some lending processes into specific categories or viewpoints taken by the mortgage lender. He looks at church property, for instance, much differently than he does income-producing property. The congregation of a church has a definite "character" that corresponds to the character or integrity of an individual borrower which is assessed by the lender as a factor in granting the loan. On the other hand, the income-producing property such as a motel is considered by the lender for its income, or "collateral" capacity. This principle can be applied to either non-profit income-producing property or a profit-making enterprise in the form of a cash return on the real estate.

This collateral security is considered by the lender in the case of the cooperative apartment with its income-producing capacity, the *long term lease.* Two separate types of cooperatives call for separate financing procedures. One is the multiple unit, sometimes a high-rise building. The other is an individual house, from a duplex to a four family unit.

From a construction standpoint, the cooperative apartment

is not necessarily architecturally identifiable as many people seem to believe; condominiums, too, are not limited to any special design. The cooperative can be of the management or the sales type. The rent is the income. Rent is controlled by the cost of the operation, which includes the mortgage loan with its interest and amortized principal placed on the mortgage. The full ownership of the cooperative building is divided among the various units so that the equity (purchase price over and above mortgage) is separately credited among the various apartments according to their relative value.

The promoters, usually a corporation, must apply for construction money based on attractive leases to be taken over by the purchasers. To be attractive, prices should be fair and reasonable with fair market value of comparable land and building costs taken into account as well as favorable interest rate and amortization terms. The prospective purchaser submits his application to the corporation representatives for a lease on the apartment of his choice and also enters into an agreement to purchase shares in the corporation.

The money to buy the land, improve it and build a home or a high rise apartment under a non-profit cooperative project, comes from the proceeds of the mortgage loan and from the members of the cooperative housing corporation. In other words, the members of the group must supply the difference between the amount of the mortgage and the total outlay of the project. See Figure 12-1, which depicts a cooperative accounting concept.

The Blanket Mortgage

Members of the cooperative do not apply for the mortgage. The corporation applies for one "blanket mortgage" covering land and buildings with the mortgage payments covering the principal, interest, taxes and insurance. Reserve accounts are also set up. The board of directors determines the amount to be contributed by each member to the payment of the mortgage, and a proportionate amount as a share of the total is allocated to each unit. Since this is a non-profit organization, any income exceeding expenses or required reserves cannot be handled as a profit to the cooperative. The surplus must be distributed at the discretion of the corporation, and it may take

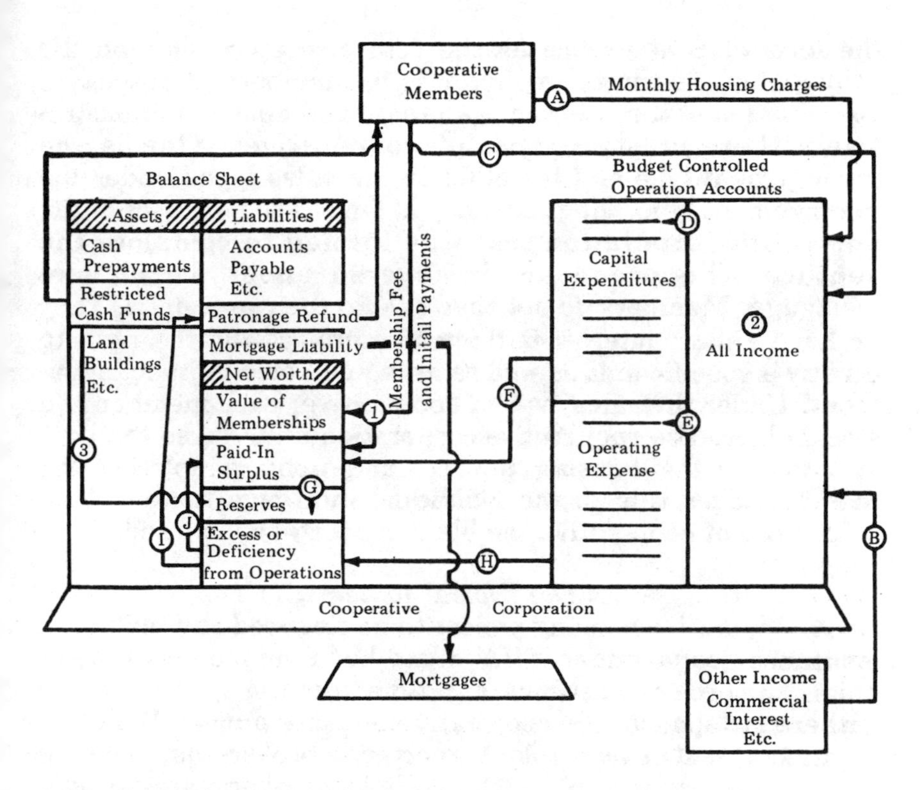

1. Membership fees are recorded in Value of Memberships Account. Any contributions in excess are recorded in Paid-in Surplus.
2. All income is recorded in a budget controlled account:
   A. Monthly housing charges from owners.
   B. Other income
   C. Cash in reserve funds reflects cash (that has been previously contributed and recorded as Paid-in Surplus) withdrawn from the funded reserves and redeposited in operations to defray an unbudgeted or periodic expense.
   D. & E. Reflect income available for capital (monthly contributions) and operating expenses.
   F. & G. Contributions are credited to Paid-in Surplus Account where they are further distributed to the reserve and restricted fund accounts.
   H., I. & J. When operations result in a surplus, it is either appropriated and credited to Paid-in Surplus or returned to the members as a patronage refund.
3. Reflects funding of reserves with contra accounts.

Cooperative accounting concept
Figure 12-1

the form of debt service for the future, prepayments on the principal of the mortgage, refunds to members or possibly a combination of these forms. As mentioned above, administratively there are two types of cooperatives. One is the management type and the other is the sales type. Under the management type, the occupancy is limited to members of the non-profit corporation that was formed to complete the venture. This occupancy is covered under the blanket mortgage. Members do not have title to their own units. They do have a share interest in the whole project and the right to occupy a specific unit as well as an equal voice in its management. Under the sales type of cooperative, each member is a stockholder. The cooperative corporation undertakes the construction of the housing project, and upon completion the member takes title to the individual mortgage thereto. In a sales type of cooperative the blanket mortgage is lifted.

The Typical Joiner

A cooperative housing project to be financed and built with insurance loans under FHA must be a minimum of eight units. The project must have eight members. A group must be gathered to sponsor the cooperative housing project. Builders, architects, real estate brokers, mortgage brokers can make up such a project group. Such skilled members are usually enthusiastic about developing cooperative housing. Veterans' groups, labor organizations, teachers' unions and other bodies also form sponsoring groups from their membership, attracting those who are familiar with real estate, construction, mortgage lending and the like. A cooperative organization formed by a sponsoring group is usually a successful procedure.

Cooperative housing laws are generally broad and permit housing cooperatives in some states, while in other states a ''limited dividend'' law provides the authority. Before soliciting members, a cooperative must become incorporated under the state law with a charter and bylaws. It must elect officers and directors. The sponsoring group arranges this plan in detail after closely analyzing its state law. In most states, however, the cooperative corporation can be formed under the general corporation laws.

Pros And Cons

There are characteristics of the cooperative which may have advantages and disadvantages to you, depending upon your situation. These include the following:

- Development of your equity.

- Income tax deductions which accrue to all owners of cooperatives.

- The right to sell or sublease subject to new buyer's or tenant's certification according to the cooperatives provisions, plus approval by the members.

- The "one landlord" concept. Conventional rentals are subject to an apartment's landlord-management relations or changes. These relations affect the rental's method of operations or the character of new tenants. This is not so with cooperatives: firm policy prevails.

- Each member's right to a voice in forming standards and in the building operation.

- Permanent occupancy not at a landlord's discretion.

The tenant owner of a cooperative unit purchases shares in the corporation that owns the entire land and buildings.

The Lender's Viewpoint

Here are some general provisions on this type of loan from the viewpoint of the lender. He must be assured that these points are covered thoroughly in the proprietary contract you and the corporation make with the individual lessee. Generally, the proprietary lease contract is an agreement made by the corporation (lessor) to sell to an individual (lessee) shares of the capital common stock sufficient in number to own a lease of a particular apartment specified in the agreement. The lessee agrees to covenants on several subjects:

- Rental amount

- Permitted uses
- Lessors' inspection rights
- Lessors' lien
- Surrender conditions at end of lease
- Condition of the premises
- Compliance with house rules
- Limits of the lessor's liability
- Mortgage agreement
- Assignment of apartment or sublease
- Lessee's termination
- General provisions

The stock issued covers the equity above the amount of the mortgage. The tenant-owner (lessee) then has no personal liability on the mortgage but must pay his proportionate share of the payments, unless he pays the full amount in cash for his cooperative apartment. He is not issued a deed. He is issued a proprietary lease. The lease entitles the stock purchaser to a long-term occupancy. In the meantime he has a restricted right to sell to a third party.

As in all mortgage loans, the lender requires the mortgagor to comply with certain conditions in making a loan on a cooperative income-producing property whether it is for profit or nonprofit. All the payments of rental, property taxes, and insurance are consolidated into the monthly payment and then paid by the mortgagee (corporation). Any transactions subsequent to the original executed mortgage are to be in compliance with the mortgage terms. Reference to these debt transactions is to be made in the body of a new or supplemental instrument. The lender reserves the right to inspect the books and accounts of the management at any time in order to determine the balance due on contracts of sale or agreements of sale on each apartment. The lender can also require a semiannual statement of the corporation or its assigns, showing the principal balance due on any or all apartments.

Depending on the security instrument of the particular state, in case of default on the mortgage or trust deed, the lender should have the right to reserve and collect all rents, profits and income from the mortgaged property. He will also be appointed the attorney-in-fact of the mortgagor. A default on a cooperative has curable features just as in other real estate security. But no portion of the mortgage will be released to any individual cooperative owner until the mortgage is paid or default is cured.

Individual federal income tax deductions can be taken on the part of the tenant's monthly payments that the corporation designates for interest and taxes. This may be possible under state law income tax procedures also. A few years ago, many cooperatives offered a two to three percent return on an investment in the equity through these income tax deductible features. Today's practice may have adjusted to current investment theory. However, a greater return is realized when comparing the "rent saving" achieved by purchase of cooperative equity with current yield in stock and bond investments where *all* interest and dividends are taxable. This is a good selling point to those who are retired and living on invested capital, as well as to young middle income families. Cooperatives of the past found favor among the rich, but you can appeal to lower income people to take advantage of the cooperative for the tax advantage. Check with your tax authority for the latest information on cooperative equity investment.

Condominiums and Joint Ownership

We have hardly mentioned the difference between the cooperative apartment and the condominium. This is because a mortgage lender is concerned not with the non-profit income-producing property vs. individual profit concepts—but simply with his recovery of the investment in case of default. Every state has its own set of laws covering both concepts as well as foreclosure procedures under either trust deed or mortgage default. The real difference between the cooperative and the condominium lies in the ownership. The cooperative is valued as a proprietary lease while the condominium is valued as a "fee estate." Both lending "securities" may be viewed by the lender as an element of risk—a philosophy that most lenders use to establish the interest rate and the loan to value ratio, as we have seen earlier. A 70% LVR has been considered the accepted risk for construction and take out loans for both kinds of property.

The condominium concept is not really new as an "own your own apartment" idea, but it grew in popularity after the war years. The postwar demand for housing saw many such projects take form in the New York, Washington D.C. and Chicago areas. Condominiums were recognized as a type of residential ownership including outright ownership of an apartment in the multiple unit building, an undivided interest in the land and common interest in the building. Although condominium building was slow in establishing mortgage lenders' interest as a real estate investment, it became an important activity when the federal government entered the picture as an insurer under FHA and HUD.

Mortgage lenders are interested in financing the individual units, but the term "condominium" does not simply mean ownership of a residential unit. It also extends to offices and utility space. Therefore, title insurance coverage must be negotiated both for the original construction loan and for the fee estate of an individual unit. While most states will not find specific legislation an absolute necessity to actually "establish" such ownership, the involvement of HUD/FHA may help resolve any questionable problems of ownership. Condominium ownership specifies individual ownership of a single apartment unit with ownership in common of the hallways, swimming pools or other facilities. The cooperative apartment, on the other hand, issues shares in the corporation along with the proprietary lease.

A clear basis of ownership must be established in a condominium. For instance, a simple terrace outside an apartment would be an easement, not an ownership, because it is part of the exterior portion of the building. For title insurance purposes the contracting surveyor must establish boundaries horizontally when he measures the elevations of the floor and the ceiling above the city datum, which every city has established. The city places an engineering benchmark in an appropriate spot within its boundaries. The city datum is a plane of a specific distance below this established benchmark. The benchmark in Chicago, Illinois, is cut on top of the bottom stone of granite base at the southeast corner of the Northern Trust Company Bank Building, which is at the northwest corner of the LaSalle and Monroe streets intersection. The Chicago city datum is a plane 17,640 feet below this benchmark. In the case of vertical boundaries, ownership may be determined by measuring interior surfaces of the perimeter walls.

Despite the difficulty of establishing the basis of ownership, the condo owner has the freedom to exercise his own financing arrangements. It is his real estate to encumber at whatever interest rate he can shop for, or at whatever LVR he might find to meet his needs. This is not so with the cooperative, with its community mortgage and its permanent term and interest rate set by the original promoter-builder until the loan is paid off. Condominium apartment owners are not liable for the defaults of other apartment owners of their mortgage. The lender would merely repossess the condo apartment. But with the cooperative this is not so, even if cash is paid for the unit over and above the mortgage in a management-type project. It is true that the sales-type cooperative is similar to the condo, but it is still limited in its scope. Owners of condos also can deduct their realty taxes and interest on their own mortgage for income tax purposes. Capital gains tax can be postponed, just like other forms of fee title ownership, if the owner reinvests in another residence within a prescribed time.

While any group of rental units comprising ordinary apartments can be converted to a condominium system, ideal plans call for construction of the original building to conform to a design that reflects obvious condo ownership. Assume that you are seriously considering the construction of a building you wish to promote as condominium housing. You realize that the deed to be delivered to each owner of a unit must legally

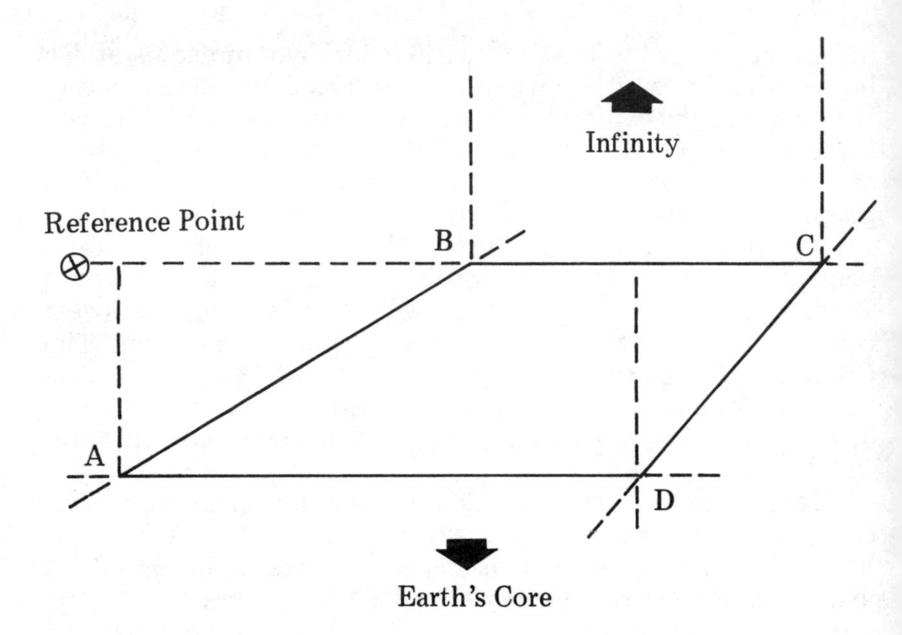

Traditional method of separating the estate
Figure 12-2

describe the portion of property to be transferred. That is,the deed must accurately describe the land and precisely where it is located. This is all that is actually required—the identification of the land and its location—as far as the traditional estate in real property is concerned. This traditional estate in real property is separated from surrounding property by stakes at each of four corners of an ordinary parcel of land, as shown in Figure 12-2. The vertical lines do not start at the top or bottom of each stake, but actually run upward to infinity and downward to the core of the earth. In between each stake a vertical plane is formed; these four vertical planes represent the estate, and partition it off from surrounding property. As in a typical suburban subdivision property can be subdivided into several smaller parcels, or "estates," as shown in the profile drawing in Figure 12-3. New vertical planes separate each estate from the other and make separate ownership possible.

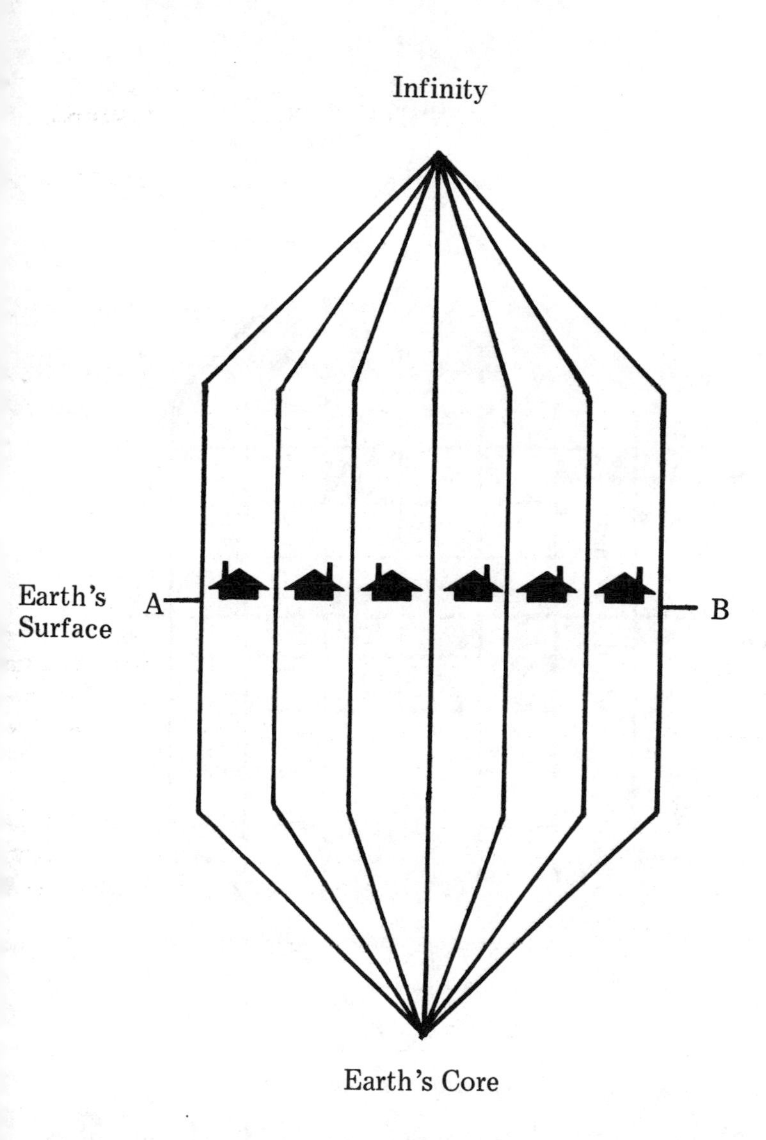

Single estates and single title from earth's core to infinity
Figure 12-3

Horizontal lines that bisect the vertical planes are also used inside the traditional estate, as shown in Figure 12-4. They prevent the vertical planes from running their course from

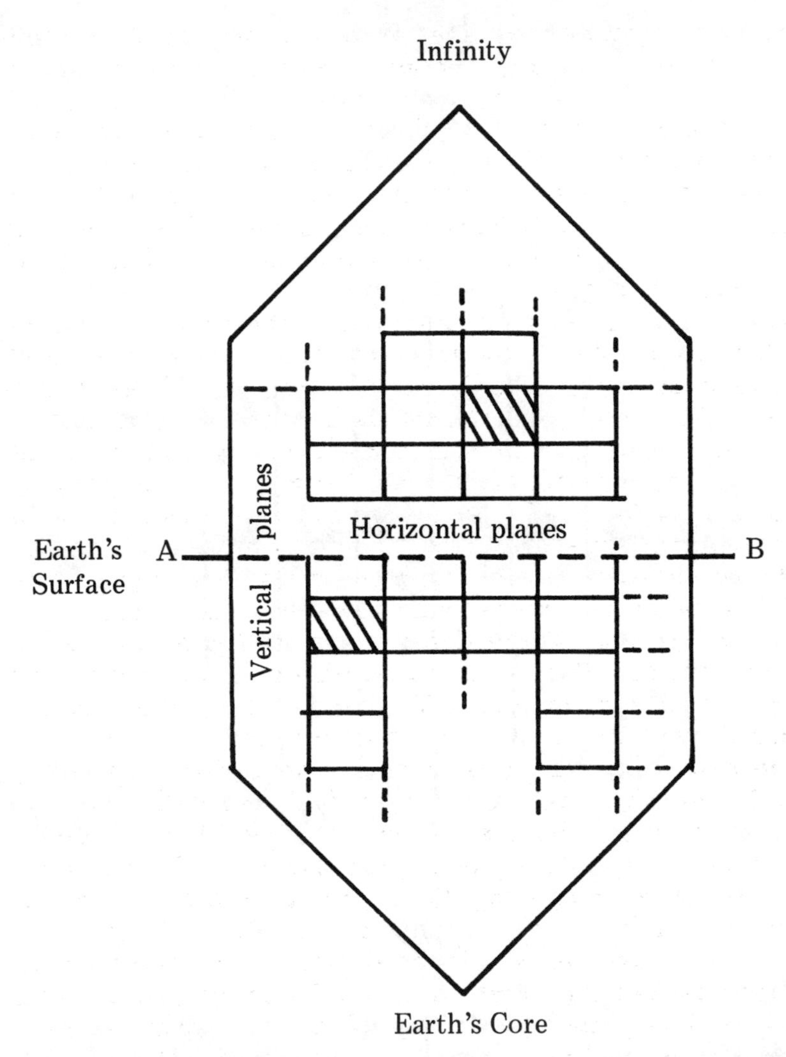

Horizontal planes form the air space estates as invisible cubes
Figure 12-4

earth's core to infinity. This forms invisible cubes of air space sometimes called "air space estates." The vertical planes are realized as they appear as walls and the horizontal planes as they appear as floors and ceilings of a dwelling unit and its space. All structures outside of the various dwelling unit spaces, as well as the land and air space outside the units, are common ownership with undivided interests. It is this common space extending outward in all directions to the boundary lines of the property that maintains the real estate characteristics of the condo units. And although the vertical planes of the unit spaces no longer extend past the ceilings and floors forming the unit, the property lines of the condominium itself do extend from the earth's core to infinity.

The plat of this condominium shows the exact location of the structure inside the common estate, and the architectural plans must show the exact location of each unit of the condo. The legal description of the boundary lines of the total property and the internal boundary walls of the dwelling spaces are set out in the enabling declaration. In other words, the unexposed surfaces of the four walls and the floors and ceilings must be in the form of a legal description. The declaration of the condominium with the plat and plans will officially come into existence when the property is recorded. The effect is to convert a previous single deed or traditional estate into a multiple number of deeded estates in the form of dwelling units within a common estate. The deed of the common estate is shared by all the individual estate owners.

Since there are common elements of the property owned by all unit owners, what about individual ownership and its relationship to the fair market value of the whole condominium? You can determine a proportion of the common elements owned by dividing the fair market value of each unit by the fair market value of the entire condominium. If a lender's appraiser uses this particular value factor as an important part of his overall valuation, he determines the unit and the building value separately. If the fair market value is $1,152,000 for the building and the fair market value of each unit is $32,000, the proportion of the common elements owned by each unit owner would be about 2.8 percent.

To repeat the distinction between these two types of property ownership, a condominium means joint ownership but not joint tenancy. An enabling declaration establishing a condominium plan refers to owners as "tenants in common" of all the remaining property. This declaration, Form 3276-A, can be obtained from your local FHA office.

13

Land Planning

The builder's objective is to develop sites for buildings so that the raw land can be utilized to its full potential, thereby justifying the initial outlay and providing top cash yield. Red tape, building codes, and environmental concerns for lower densities have driven up the cost of land per unit. Planning ingenuity is therefore a continuing process rather than a single action or an unchanging construction design. The lender's objective is to invest community dollars, safeguarding all savings or contributions entrusted to his care. He hopes for the highest yield his portfolio can bear. Keen competition dictates that long-term security of the loan depend upon the quality of the development. The lender's planning ingenuity is also a continuing process rather than a single action, as he approves loans and disburses funds over a period of time to many projects in his area. The public planning agency provides general plans, programs and appropriate regulations for the orderly development of the land areas under its jurisdiction. Planning jurisdictions of public agencies sometimes overlap; often local controversies in zoning and planning are related to maintaining public health, safety and welfare, with each agency setting up its own zoning and planning criteria for those ends.

The builder-developer works with two assistants, both indispensible—the financial backer and the public agency. They give advice on the loan and on technical procedures of the land plan.

Basic to any land planning is a system of identification of the land. In your dealing with these people, you must refer to your proposed land development as a legal description on a Township Survey System map.

The Legal Description

A property is usually identified with a mailing address before a technical legal description is used. But if you have a proposal to place improvements on country acreage, there are no addresses yet. It is too early to hire surveyors for the proposed construction site; so unless you already own the land you can easily obtain the legal description by checking the ownership lists at the assessor's office for a preliminary determination of the boundaries. The rectangular survey system is probably the most widely used, as about 60% of the states use it. This system originated with the Continental Congress, and various changes occurred in it between Jefferson's time and the year 1796. In that year the township was established as a unit of land measurement with 36 sections, each section one mile square. Horizontal distances are expressed in miles, chains or links. Meridians are lines running north and south, and base lines east and west. These are imaginary references used in the rectangular survey system that have nothing to do with latitude or longitude lines used in chartmaking for navigational purposes. The United States Surveyor General has jurisdiction over the surveys of all public lands. In order to locate and describe land, certain base lines and meridians are given systematically descriptive names. From these base lines and meridians the location of land may be accurately determined.

The public domain is divided by north and south lines six miles apart called "ranges," and east and west lines also six miles apart called "township" lines. Squares formed by the intersection of these "ranges" and "township" lines are called townships. The standard township is six miles square and contains 36 square miles. Specific land measurements, their descriptions and conversions, can be found in Appendix XII. The intersection of the base line and the meridian is the starting point of calculations east or west, north or south, to locate a definite township. Ranges are numbered east or west from a principal meridian, while townships are numbered north or south from the principal base line. Thus, township 4 north, range 3 east, would be three townships to the east (to the right) of the principal meridian and four townships to the north (up) of the principal base line. Township 5 south, range 4 west, would be five townships south (down) of the principal base line and four townships west (to the left) of the principal meridian.

The locations of the various principle meridians are calculated from the first north-south meridian, known as Ellicott's Line. It is located between the states of Ohio and Pennsylvania. Principal meridians are located in Indiana, Illinois, Wisconsin, Arizona, California, Utah and many other states. Some principal meridians are numbered while some are referred to by name specifically, such as the Boise Meridian of Idaho, or the Third Principal Meridian of Illinois. California has three meridians, the Humboldt, the Mt. Diablo, and the San Bernardino, located in the northwest, the central, and the southern part of the state.

Townships are further divided into sections. Each township contains 36 sections. Each section is one mile square and therefore contains one square mile of land. The township is six sections long (six miles) and six sections wide (six miles) and therefore contains 36 square miles. A section may then be divided for more specific description into quarter sections and fractions of quarter sections. A quarter of a quarter section is the smallest division recognized by the statutes. Figure 13-1 shows a township plat, the numbers running from 1 to 36. Note that section number 15 is divided into quarters of quarter sections, each small square representing the location of a 40 acre plot of land. Figure 13-2 shows one of the sections in which a ten acre plot of land is depicted as a shaded area.

In a legal description the tract, which is a subdivision of the hypothetical ten acres in our example, is further identified by block and then by lot. A legal description used in locating this parcel of land and the specific lot in question is as follows:

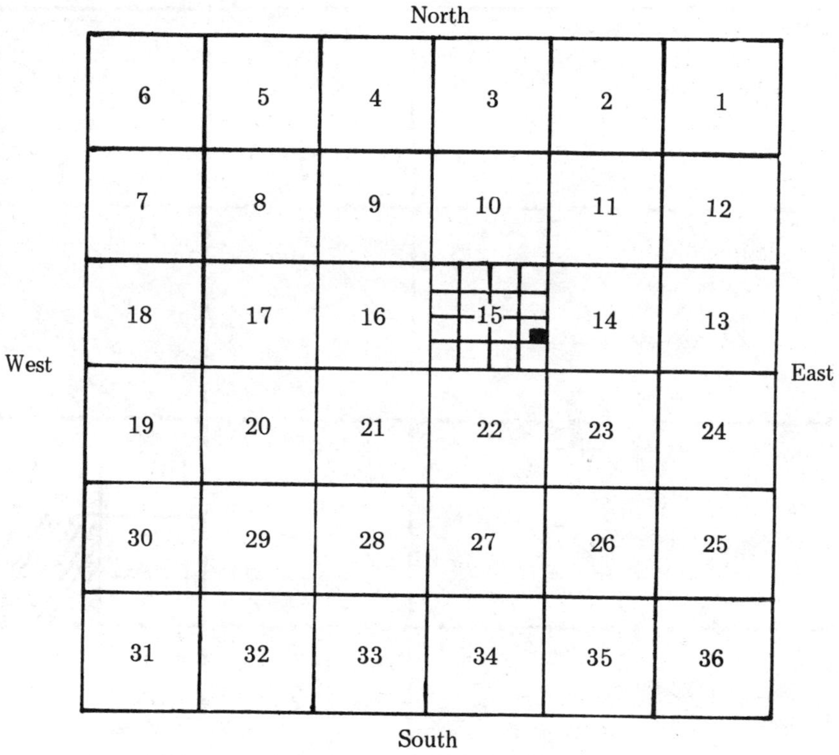

North

West							East
	6	5	4	3	2	1	
	7	8	9	10	11	12	
	18	17	16	15	14	13	
	19	20	21	22	23	24	
	30	29	28	27	26	25	
	31	32	33	34	35	36	

South

Township plat of 36 square miles

Figure 13-1

> **Lot 14, Block 7, Parkview tract, being a subdivision of the SE ¼ of the NE ¼ of the SE ¼ of Section 15, all in Township 7 South, Range 13 East of the San Bernardino Meridian.**

Planning for the total urban area is a complex matter. It has been broken down to three levels—metropolitan, community, and neighborhood planning. *The metropolitan planner* or area wide financial backer, as well as the potential developer, is

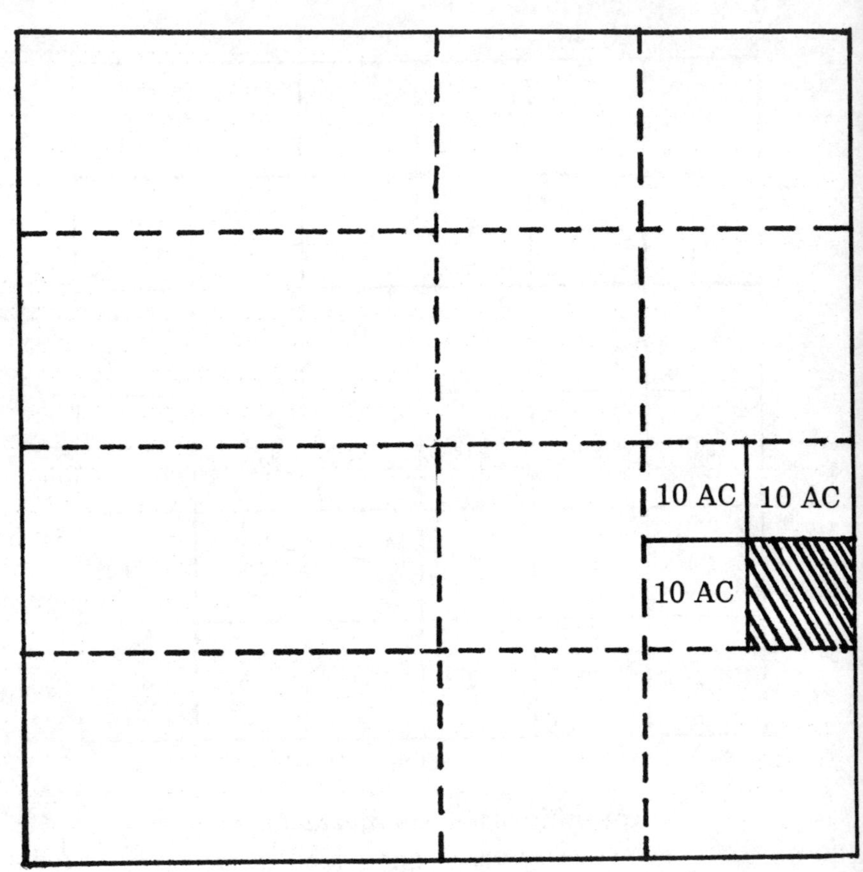

SE ¼ of NE ¼ of SE ¼ of section 15, ¼ of ¼ of 160 acres
Figure 13-2

concerned with the economic base of any city of at least 50,000 inhabitants. Most established metropolitan areas have long ago coordinated the activities of their various political jurisdictions. Your planning contacts are concerned with major land uses, population distribution and economic activities. These activities include water supply, sewage, air pollution control, major natural drainage and regional open space. For the developers, these planning factors come down to basic feasibility considerations such as employment, the housing market, freeway locations and the land use intensity. Lenders are not the specialists in your general planning efforts, but they are usually up to date in general planning information. You can start your research with the metropolitan planning agency, which serves as a data bank and clearing house for general planning information available to the public.

The *community planner* works with an area of sufficient population to support a high school, community center with a variety of shops, stores, recreational establishments, business and professional offices and post office. The area may be included in a group of local community units into which a large public agency has divided a more populated city or a large unincorporated area for study, planning, and administration of land use and common facilities. The population here is from 10,000 to 50,000. A transportation network interconnecting with a metropolitan area is desirable. Provision may be available for an official map, subdivision regulations, and a zoning ordinance.

The *neighborhood planner* thinks in terms of homesites and schools. A playground should be located near the school. The internal street system should discourage through traffic and protect against undesirable traffic patterns. The neighborhood is a residential and commercial unit within a very small land area. This may be in a community with a high land use intensity near the center of a large city, or in an outlying area in a community of very low land use intensity. The general plan for a neighborhood identifies boundaries showing the land use pattern with the location and size of the major streets, interior collector streets and common facility sites. The neighborhood usually has well defined boundaries such as arterial streets, expressways, rivers, railroads, major industrial areas or other large non-residential uses.

Generally, the overall planned scheme involves some slight increases in utility costs as compared with an unplanned or incomplete scheme, such as added costs in curved curbing. But in the long run these costs are far outweighed by the savings of the overall plan. Although the developer must conserve in order to profit, in the end the greatest saving realized through proper planning and control lies in the avoidance of certain difficulties and expenses that the property owners, taxpayers and even the lenders may have to struggle with in later years. Adequate community needs should be provided for early in the original plan. An example of this is the overcrowding and the inadequate sewer and water facilities which are creating a major crisis for many communities. This is the result of little or no land planning early in the development. Communities throughout the country have suffered tremendous losses in community funds to install sewers, simply because the original septic tank system had been installed in areas of unfavorable soil conditions. These septic tank systems, the result of little or no planning, later become serious community health hazards. Many studies have shown that with proper planning, much money can be saved in street improvement costs, and several acres of land added to the area set aside for recreation and other purposes, with the same net return to the developer.

To the financial backer and the local governing authority the developer has a responsibility to the community beyond simply producing a profitable investment. He contributes to the creation of a new neighborhood or community, which will be greatly influenced by the character and standard of his development. Practical experience shows that any planned unit development or cluster housing project should be located with a correct reference to the city plan, population trends, business and industrial locations, transportation and other facilities and available utilities. Logical city growth should be obvious to the potential subdivider, especially where local governments have prepared a master plan for the city, county or region.

The land plan is an advantage to you, your financial backer, and the public agency in the following ways:

- The plan secures the investment for generations while preserving the tax base through protection of property values and property rights.

- Good planning that respects the contours of the land holds down development costs by requiring fewer lineal feet of streets and curbing plus less cutting and grading.

- The plan helps to avoid the misuse of property and can even prevent damage to the health, safety and convenience of the community as a whole.

- Planning reveals in advance hazardous or unsatisfactory conditions to be avoided, thereby eliminating future depressed property values.

- A good plan can save municipal governments unnecessary expense by providing a suitable site for schools, parks or appropriate recreational facilities at a time when land value is low.

A cluster subdivision of detached homes amid preserved open space is one widespread example of better community environment through planned unit development at moderate suburban densities. In Radburn, New Jersey, the municipality uses this technique in the super-block pattern. Another example is the townhouse development featuring landscaped malls and recreational facilities.

A lender thinks in terms of the overall neighborhood as the best security for an individual project. Its stake in home mortgages, for example, is as great in the neighborhood as it is in any specific property. In the case of subdivisions, the long term security of the lender's investment rests first on the quality of the whole tract development. That is why the savings association or financial backer, as one of your assistants, must be sure he will not have to live with your construction and development mistakes. He is an investor, not a gambler. Here are the major investment criteria used by your financial backer.

- There should be a study of housing demand and growth trends for the city relative to other potential sites to back up your site selection.

- Sewers, water, and power systems availability and convenience considerations are important to him. Central systems are always best. Avoid power poles in the streets.

- Streets and block patterns should observe the natural topography, thereby holding down grading to a minimum.

- Lot sizes must be adequate but with economical shapes. All individual units or dwellings should be sited on the lots for attractiveness.

- Neighborhood considerations should include recreation spaces, sites for schools and churches, with special emphasis on existing covenants.

Planning For Compliance

By compliance is meant following the general plans of the zoning ordinance, or carrying out procedures for maintaining public health, safety and welfare. Local regulations limit the purposes for which land may be used and the maximum intensities of land use. The local agency will actually issue guidelines for you. You must comply with these local guidelines and the existing zoning regulations, if there are any. Some large municipalities such as Houston, Texas, have never seen the necessity for zoning. But your compliance with local zoning ordinances gives considerable protection to property values by creating and maintaining desirable community conditions.

But sometimes a zoning ordinance may permit business, industrial and residential uses at densities and extents which are greatly in excess of the probable long-term demand. The local program for land subdivision which is regulated by such zoning ordinance should be based on probable long term demand rather than on the most intensive use permitted under a zoning ordinance. The subdivision program, therefore, frequently provides for a less intensive land use than that designated on a zoning map. A zoning ordinance usually permits the development of a parcel of land for a designated use or for another less intensive use.

Land Use Intensity and Density

These terms are not the same, though they are easily confused. Land use intensity (LUI) means the overall structure mass to open space ratio in a single piece of developed property. It is a correlation of the amount of floor area, open space, livable space, recreation space and car storage space of a property with the size of its site or land area. *Density,* on the other hand, is the number of living units or people per unit of land area. An unwarranted low land-use intensity number tends to affect any multi-unit project adversely through underuse of the land. Assigning this number is a crucial step in the preliminary consideration of any multiple housing proposal. A correctly assigned number establishes a workable basis for the planning, construction and operation of a successful housing project. This success is measured both by market absorption and by long-term values, whether the project is for rental or for home sales. However, an unwarranted high land-use intensity number may affect your project adversely. It can lower marketability and livability standards. Where local governments have prepared a master plan of the city, county or region, look for this land use intensity rating in your feasibility study, as it is important both to the master plan and to your development. The most desirable conditions to be sought in land use intensity are outlined in HUD Handbook 4140.1 "Land Planning Principles for Home Mortgage Insurance," May 1973, available at your local HUD office.

Land use intensity is expressed as a number: the land use intensity, or LUI, number. Of two intensity numbers, the higher can have the more detrimental effect on the project because an overly high intensity lowers the comfort, convenience and appeal of the entire project. This is reflected in lower market appeal or lower rentals. Some developers try to build with high land use intensity in mind with the idea of increasing profits. This is a fallacy, as the rest of this chapter will show. From the development of area data leading to a land use intensity number to the finished house on the market, try to build with a land use intensity number which is slightly lower rather than higher.

Establishing Land Use Intensity Numbers

The land use findings and recommendations as presented to HUD-FHA are gathered and analyzed by appraisers and land planners.Your project may not be covered by loans involving HUD-FHA insured mortgages; but in some cases a market analyst, an architect, a site engineer and a sanitary engineer would participate. Their written findings and recommendations regarding the intensity number for the site are given to the chief underwriter of an insuring office, or to the assistant director for single family mortgage insurance (ADSF/CU) in the case of an area office. After considering the recommendations of his staff, he determines the land use intensity number.

The land-use intensity number of a housing site varies substantially with the site's location in the community's space pattern. This is true even in a relatively small community. The highest intensity is in the center of the community and the lowest is generally in the outlying area. In the transitional areas between the two are intermediate intensities. The site may also be located along a corridor connecting the center to another urban area. In determining the land use intensity of a proposed site, then, you must understand the present and prospective space arrangements of the land use pattern of the community and position your proposed site in that land use arrangement. See Figure 13-3.

You can obtain data related to land economics from your local planning commission. Statistics for making intelligent land use ratings are available from surveys and forecasts of industrial and other economic activity, population size and anticipated growth. From this data you can make general conclusions as to the total range of intensities appropriate to the land in the community currently and for the foreseeable future. Take into consideration the present size, growth rate and probable ultimate size of the community. A small community with a slow growth rate and an ultimate size that is also relatively small will have a much narrower and lower range of land use intensity ratings than a larger community. For instance, an isolated town with a population of less than 2,000 and slow growth would have a much lower upper limit to its range of land-use intensities than a well-located, rapidly-developing suburb in a metropolitan area.

On the other hand, a community with an expanding pattern sometimes presents difficult problems in determining a proper

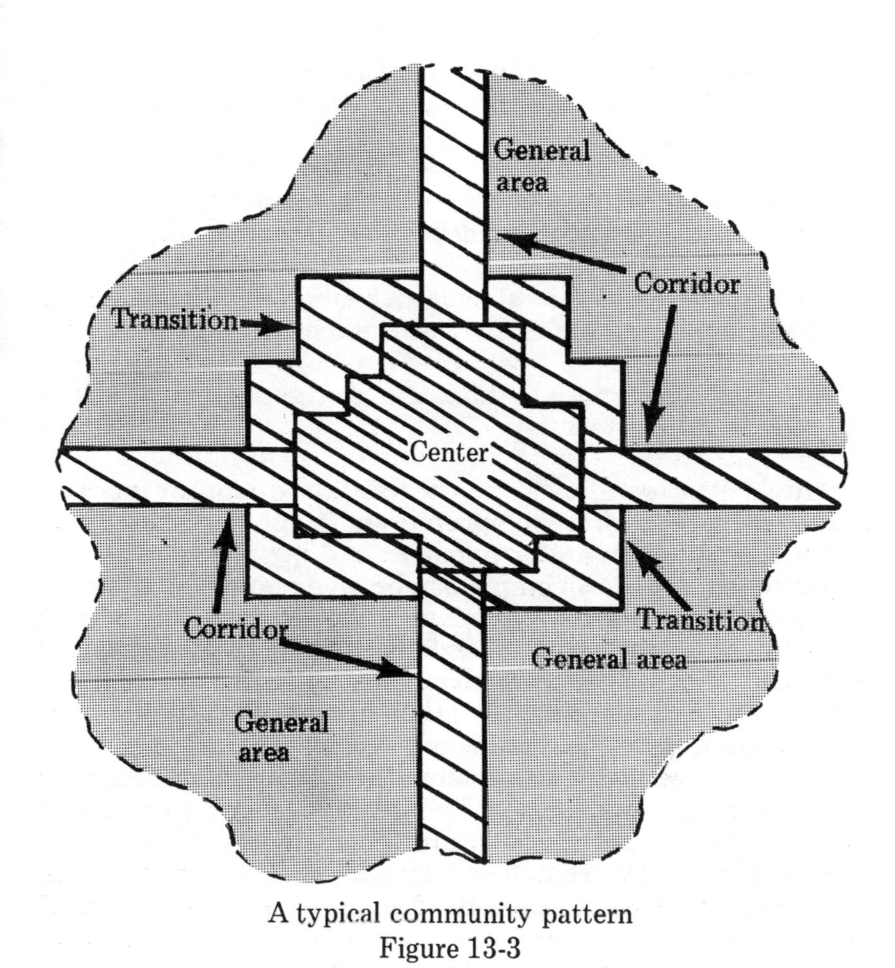

A typical community pattern
Figure 13-3

rating of land-use intensity for its satellite developments in fringe areas. In these cases you must forecast the emerging and future intensity pattern. Many fast-developing residential areas follow a normal and predictable pattern. New stimuli or changed conditions, however, may warrant more intensive land use in the future development pattern. Where this is the case, you can build your projects with a higher land-use intensity than is current in the area. To determine the degree of higher land use intensity you can count on for building in these areas, study the space arrangement and the land intensity range in conjunction with the following:

- A thorough study of the probable effect higher intensities would have on existing development in the area and in adjacent undeveloped areas

- Consultations with planning, zoning and engineering officials of the municipal government

- Consultations with realtors and mortgage lenders active in the area

These studies should be thoroughly documented and placed on file with all facts, opinions and judgements justifying a higher land-use intensity number.

You should make a comparison between land-use intensity and density in order to thoroughly understand the overall community pattern. Land use-intensity (LUI) is less variable. The term "density" as commonly used in land use planning, zoning and site planning means the number of living units per unit of land area, or the number of people per unit of land area. Because there is a wide variation in the size of living units and the number of occupants of living units of any given size, density is a rather crude measure of the degree of land use. This is why HUD/FHA does not use density directly as a land-use measure in its multi-family minimum property standards.

Land-use intensity is more reliable than density. The concept of land-use intensity as used here starts with and is directly related to *floor area ratio,* which is the amount of floor area to the amount of land area. Thus, it is a measure of the total permitted floor area on a site of a given size. In residential property, total floor area roughly determines the number of living units and the number of people. Thus, land-use intensity numbers based on floor area and land area have some general relationships to density based on living units and land area. Because density is unresponsive to wide variables in living unit size and household size, it can be compared to land-use intensity only in general terms and at the risk of gross

Land-use Intensity (LUI)	Floor Area Ratio (FAR)	Floor Area (sq.ft.) per Gross Acre	Density in L.U. per Gross Acre	
			1089 s.f./L.U.	871.2 s.f./L.U.
0.0	0.0125	544.5	.5	.625
1.0	.025	1089	1.0	1.25
2.0	.05	2178	2.0	2.5
3.0	0.1	4356	4.0	5.0
4.0	0.2	8712	8.0	10.0
5.0	0.4	17424	16.0	20.0
6.0	0.8	34848	32.0	40.0
7.0	1.6	69696	64.0	80.0
8.0	3.2	139392	128.0	160.0

Land-Use Intensity Related to Floor Area and Density
Figure 13-4

misinterpretation. A valid comparison can be made, however, if the living unit size or the household size (the variable in density) is kept constant at some specific size.

Land Use Intensity Scale

The floor area to land area, a rate or measure, is calculated from a measurement scale like that found in Figure 13-4. For land-use intensity, the scale is based first and most directly on the relationship of total floor area to total land area, that is, on the floor area ratio.

The *floor area ratio* (FAR) scale appears in the second column of Figure 13-4, a chart of land-use intensity related to floor area and density. The chart starts at a floor area ratio (FAR) of 0.0125. This is arbitrarily called 0.0 on the scale of land-use intensity (LUI). Since the floor area ratio multiplied by the land area (LA) equals the floor area (FA), then FAR 0.0125 times the 43,560 square feet per acre equals 544.5 square feet of floor area per acre at land use intensity 0.0. To visualize LUI 0.0, this floor area per acre can be converted to living units per gross acre. For example, 544.5 square feet per acre equals one modest home of 1,089 square feet floor area

(2 x 544.5) with 2 acres of land. This is shown on the first line of Figure 13-4. Each full unit on the intensity scale is equal to a 100% increase in the floor area ratio. This increases the floor area 100% as related to the land area. This doubles the number of living units of any given size for each full unit on the intensity scale.

This doubling on the intensity scale and its origin at LUI 0.0 with FAR 0.0125 makes it easy to remember and visualize the entire scale. Note in line 2, columns 3 and 4, that LUI 1.0 means 1,089 square feet of floor area or 1 house of modest size per acre of land. Remember that LUI 1.0 equals 1 LU per 1 acre for a house of a little over 1,000 square feet (1,089). Remember the doubling of the intensity scale. It is easy to remember this little table which visualizes the scale:

LUI	LU/ acre for 1,089 sq. ft. LUs
1	1
2	2
3	4
4	8
5	16
6	32
7	64
8	128

Memorizing another point on the scale is also helpful. LUI 3.0 is FAR 0.1. Starting from this, any FAR value is easily found mentally by doubling:

LUI	FAR
3.0	.1
4.0	.2
5.0	.4
6.0	.8
7.0	1.6
8.0	3.2

Gross Land Area—Another Base

As noted earlier HUD/FHA does not use density directly as a measure of intensity. But it will be helpful to compare the land base of density, as a preliminary to considering land as a base in the identification of the intensity number. *Gross density* refers to all the land used by the development, including on-site streets and one-half of the bordering streets. This is the gross land area or gross acreage; the result is gross density. In net density calculations, street area is not counted in the land area.

Sometimes the public streets in the area are not counted when calculating net density, but privately owned streets are counted. The result is another variety of net density, an ambivalent measure of doubtful meaning. In recent years HUD/FHA has used the latter variety of net density. Do not be confused when paved parking areas in stub compounds are counted, as opposed to bays along streets which are not.

The HUD/FHA multi-family minimum property standards presently take gross land area as the firm base for land-use intensity. The intensity scale refers to the gross acres of all land benefiting a project. Within certain limits for land area, one-half of any abutting street right-of-way is included in the gross acreage. Also included in the land area for the intensity number is one-half of any abutting park, river or other beneficial open space which has reasonable expectancy of perpetuity. Because the gross acreage of total land area (all the area benefiting the project) is used, the land-use intensity measures are more realistic and more reliable than data based on net acres.

Net and Gross Area Conversions

If existing information based on *net area* is used in a site analysis, it is necessary to convert it to gross area for use in comparisons with land-use intensity numbers. Net area customarily excludes all street rights-of-way. These usually range from fifteen to thirty percent of the total land area.

Convert by subtracting the percentage of street area from 100% and dividing the net area by the difference. For example, knowing that the net area is 100 net acres and the street is 20% of the gross area, what is the gross area? The answer is: 100 net acres ÷ (100%-20%) = 125 gross acres.

On the other hand, gross area converts to net area by subtracting the percentage of street area from 100%, and multiplying the gross area by the difference. For example, knowing that the gross area is 125 gross acres and the street area is 20% of the gross area, what is the net area? The answer is: (100%-20%) x 125 gross acres = 100 net acres.

Net and Gross Density Conversions

Gross density converts to net density by dividing the number of living units per gross acre by the difference between 100% and the percent of street area. For example, knowing that there are four living units per gross area and the street area is 20% of the gross area, how many living units are there per net acre? The answer is: 4÷ (100%-20%) = 5 living units per net acre.

Net density converts into gross density by multiplying the number of living units per net acre by the difference between 100% and the percent of street area. For example, knowing that there are 5 living units per net acre and that the streets are 20% of the gross area, how many living units are there per gross acre? The answer is: 5 x (100%-20%) = 4 living units per gross acre.

Other Elements of the LUI Scale

As mentioned earlier, the land-use intensity number not only relates floor area to land area but also correlates other planning elements. These other elements are open space, livable space, recreation space, occupant car storage and total car storage including guest cars.

Open Space Ratios Open space, livability space and recreation space are expressed in the intensity rating system in terms of the open space ratio (OSR), the livability space ratio (LSR), and the recreation space ratio (RSR). The ratio is simply the total area of the open space, livable (outside non-vehicular) space or large recreation space divided by the total floor area.

Car Storage In the LUI system, car storage capacity is expressed in terms of car ratio, which is the number of cars to be stored divided by the number of living units.

We have seen that the floor area ratio as a base for the intensity scale has a very simple mathematical relationship to the LUI—it doubles with each LUI unit. But these other elements of the scale have a more complex relationship to the determination of land-use intensity and your planning department can help you with these. If you need a more thorough analysis, the HUD Handbook 4140.1 mentioned above covers these subjects in the event your own land intensity studies justify further preparation.

How Many Living Units?

LUI related to floor area In Figure 13-4 and the preceding discussion, we have seen that the scale for land-use intensity is directly related to floor area ratio and to floor area; also that the number of living units per acre is directly related to an intensity number because the size of living units varies. Columns 4 and 5 of Figure 13-4 show the great magnitude of this variation for living units of 1,089 and 871.2 square feet.

LUI related to open space We have also seen that the land-use intensity number is related to the amount of open space of several kinds, namely, livability space, recreation space, as well as total open space.

LUI range It becomes clear, then, that the land-use intensity scale is a measure of overall relationships of structural mass and open space. In the intensity determination these complex relationships are distilled into a single numerical scale. Grossly simplified, the intensity scale may be said to run: (1) from LUI 0.0 for very low rural-type land use through LUI 3.0 for suburban type land use with about 4 modest houses per gross acre, (2) to LUI 8.0 for very intensive urban land-use for high-rise apartments with 82 to 348 LU per acre, depending on the unit size and the street percentage.

Each number and decimal on the scale translates into a particular level of land-use intensity based on the maximum floor area to the land area and on the minimum of open space, livable space and recreational space for the floor area.

The Sponsor's View

A natural question at this point is, "How does the sponsor of a proposed development find out how many living units he can get on his site?" If it is a HUD/FHA development, the sponsor determines the number of living units acceptable to HUD/FHA on the site by (1) obtaining the land-use intensity number of the site from HUD/FHA, and (2) preparing a

	A	B	B compared to A	
			Difference	Percent
Land-use intensity number	4.0	5.0	+1.0	...
Living units per gross acre	8	16	+8	200
Outdoor nonvehicular livability space in sq. ft. living unit	2,831	1,198	-1,633	42
Raw land cost per living unit	$ 2,000	$ 1,000	-$1,000	50
Total cost per living unit*	$20,000	$19,000	-$1,000	95
FHA 203b home loan:**				
Down payment	$ 950	$ 850	-$ 100	90
Monthly payment	$ 134.77	$ 128.41	-$ 6.36	95

***Assumes $18,000 per living unit as the cost for building, land improvements and other costs, except raw land costs.**

****Based on 30 year mortgage at 7% interest plus ½% FHA premium.**

Developer A with a lower intensity offers a property with more than double the living space of B's property with only $100 increase in the down payment and only 5% ($6.36) larger monthly payment than B. Developer A offers a better buy and probably outsells B.

Fallacy of High Density
Figure 13-5

project planning program and project design properly related to the assigned site number. An illustration of a simple approach to developing a planning program for a project, starting with the intensity number, is in FHA booklet 2600 (HUD 4910.1), "Minimum Property Standards for Multifamily Housing." HUD/FHA provides blank FHA Forms 1028 and 1029 of the method illustrated here, for the use of sponsors and their design professionals.

Effect of Higher Intensity

A more important question is, "How would a higher land-use intensity affect the financial soundness of the proposed project?" This was briefly mentioned earlier where it was stated that high intensity has a more detrimental effect because overly high intensity lowers the comfort, convenience and appeal of the entire project, which is reflected in lower market appeal or lower rentals. Too high a land-use intensity is false economy. Consider the case of Developers A and B in Figure 13-5. Each developer has a land cost of $16,000 per acre, and $18,000 for all other costs to produce 1089 square foot townhouses on similar tracts. Developer A develops his property at a land-use intensity of 4.0; Developer B develops his property at a land-use intensity of 5.0. The results are shown in Figure 13-5.

Since raw land is a low percentage of the total housing cost, great reductions in total package costs cannot come from your "squeezing" the land. We have seen that intensification of land-use produces relatively small cost savings at the sacrifice of very great losses in livable space. As livability affects marketability and rentability, it is clear that inappropriate intensification of land use can be financially hazardous. In determining the intensity for a specific site, make an economic comparison like the type illustrated above, using applicable local data.

14

Sponsor's Guide To Government Programs

This chapter will familiarize you with the federal government's participation in public housing and its relationship to private enterprise. Whether or not your sponsorship is by government mortgage insurance or guarantee, you can easily obtain all the "know how" information you need. Publications are prepared for use by the lenders, borrowers, buyers, sellers, speculators, builders, contractors, developers, subdividers, brokers, architects or any potential sponsor of any real estate venture. In fact, there are many constituent housing bureaus within the executive branch of the government willing to provide you with the building and planning information. We can cover the main programs here which affect any possible sponsorship. Also see Appendix XIV.

HUD And The Builder

The government does not make building or housing loans; financial institutions do. Money is never loaned by HUD/FHA, nor does it build homes. The purpose of HUD is to assist families in financing homes and to stimulate investment in the construction industry. A very small equity, or down payment, is required to build or purchase while following its guidelines on a firm basis.

The FHA Mortgage Insurance lowers the risk to the lender. He can therefore offer better terms to the borrower than would ordinarily be available in conventional loans. The terms, however, cannot vary from the requirements established by HUD/FHA.

We have tried here to present the main points that are specific to the lending procedure under HUD/FHA where the financing pertains to a particular type of property. The following digest of insurable loans covers property as discussed throughout the book. Not included here is the profit-making motel structure, which would not be under a government program that insures or guarantees mortgages. Information on the availability of the complete handbooks is found in Appendices XIV and XV.

203(b) Insured Mortgage on New Construction or Existing Structure

a. *Purpose of Mortgage:* The purpose of the mortgage is to finance the acquisition of proposed, under construction, or existing one-to four-family housing, or to refinance existing indebtedness on such property.

b. *Amount Insurable:*

(1) Occupant Mortgagor:

(a) $67,500 one-family.
(b) $76,000 two-family.
(c) $92,000 three-family.
(d) $107,000 four-family.

(2) Non-occupant Mortgagor: 85% of owner-occupant amount, except under special escrow commitment procedures.

c. *Loan-to-Value Ratio:*

(1) Occupant Mortgagor:

(a) Approved by HUD/FHA prior to the beginning of construction or completed more than one year on the date of application:

97% of first $25,000 of sum of the HUD/FHA estimate of property value and closing costs, plus 95% of the next full amount of HUD/FHA insurance, of the appraised value and closing costs.

If the mortgagor is a veteran (Single-family only):

100% of the first $25,000 of sum of the HUD/FHA estimate of property value and closing costs, or $25,000 plus prepaid expenses less $200, whichever is less, plus 95% of the full amount of the appraised value and closing costs up to the mortgage limits.

(b) Construction begun or completed less than one year on the date of application; HUD/FHA limits the mortgage amount as follows:

(2) Non-occupant Mortgagor: 85% of the owner-occupant (non-veteran) amount, except under special escrow commitment procedures.

d. *Term of Mortgage:*

(1) 30 years or ¾ of the remaining economic life, whichever is less. The 30-year limit may be extended to 35 years if the mortgagor is unacceptable for the 30-year term and the property was constructed subject to HUD/FHA or VA inspection.

(2) Operative Builder: 20 years or ¾ of remaining economic life, whichever is less.

e. *Interest Rate:* Current interest rate as determined by the

Secretary of Housing and Urban Development.

f. *Insurance Premium:* ½% on declining scheduled balances.

g. *Initial Service Charge:*

(1) $20 or 1%, whichever is greater, or

(2) $50 or 2½%, whichever is greater, if the mortgagee makes partial disbursements and inspects construction.

h. *Application Fee:*

(1) $65 proposed.

(2) $50 existing.

(3) $15 if a VA Certificate of Reasonable Value accompanies the application.

i. *Special Factors:*

(1) Certification that the mortgagor received a HUD/FHA statement of appraised value required on one or two-family dwellings.

(2) A builder's warranty is required on proposed construction.

(3) Eligible for open-end advances where local law permits.

(4) A mortgagor 60 years of age or older may borrow the downpayment, settlement costs, and prepaid expenses from an approved corporation or individual.

(5) The VA Certificate of Reasonable Value may be presented to establish the value of property.

Section 203(h) Home Mortgage Insurance For Disaster Victims

a. *Purpose of Mortgage:* The purpose of the mortgage is to finance the acquisition of proposed, under construction, or existing one-family housing by an occupant-mortgagor who is a victim of a major disaster.

b. Amount Insurable: $14,400

c. Loan-to-Value Ratio: 100% of sum of the HUD/FHA estimate of property value and closing costs.

d. *Term of Mortgage:* 30 years or ¾ of the remaining economic life, whichever is less. The 30-year limit may be extended to 35 years if the mortgagor is unacceptable for a 30-year term and the property was constructed subject to HUD/FHA or VA inspection.

e. *Interest Rate:* Current interest rate as determined by the Secretary of Housing and Urban Development.

f. *Insurance Premium:* ½% on declining scheduled balances.

g. *Initial Service Charge:*

(1) $20 or 1%, whichever is greater, or

(2) $50 or 2½%, whichever is greater, if the mortgagee makes partial disbursements and inspects construction.

h. *Application Fee:*

(1) $65 proposed.

(2) $50 existing.

(3) $15 if a VA Certificate of Reasonable Value accompanies the application.

i. *Special Factors:*

(1) Certification that the mortgagor received the HUD/FHA statement of appraised value required.

(2) A builder's warranty is required on proposed construction.

(3) Eligible for open-end advances where local law permits.

(4) The VA Certificate of Reasonable Value may be presented to establish the value of property.

Section 203(i) Home Mortgage Insurance For Outlying Area Properties

a. *Purpose of Mortgage:* The purpose of the mortgage is to finance the acquisition of proposed, under construction, or existing one-family nonfarm housing or farm housing on 2.5 or more acres adjacent to an all-weather public road.

b. *Amount Insurable:* (housing credit shortage areas)

(1) $50,625 occupant Mortgagor.

(2) $43,000 or 85% of occupant/mortgagor amount. Operative builder or nonoccupant mortgagor.

c. *Loan-to-Value Ratio:*

(1) Occupant Mortgagor:

(a) Approved by HUD/FHA prior to beginning of construction or completed more than one year on date of application: 97% of sum of the HUD/FHA estimate of property value and closing costs.

(b) Construction begun or completed less than one year on date of application: 90% of sum of the HUD/FHA estimate of property value and closing costs.

(2) Operative Builder or nonoccupant mortgagor: 80% of sum of the HUD/FHA estimate of property value and closing costs, or $36,000, whichever is less.

d. *Term of Mortgage:*

(1) 30 years or ¾ of the remaining economic life, whichever is less. The 30-year limit may be extended to 35 years if the mortgagor is unacceptable for 30-year term and the property was constructed subject to HUD/FHA or VA inspection.

(2) Operative Builder: 20 years or ¾ of remaining economic life, whichever is less.

e. *Interest Rate:* Current interest rate as determined by the Secretary of Housing and Urban Development.

f. *Insurance Premium:* ½% on declining scheduled balances.

g. *Initial Service Charge:*

(1) $20 or 1%, whichever is greater, or

(2) $50 or 2½%, whichever is greater, if mortgagee makes partial disbursements and inspects construction.

h. *Application Fee:*

(1) $65 proposed.

(2) $50 existing.

(3) $15 if a VA Certificate of Reasonable Value accompanies the application.

i. *Special Factors:*

(1) Certification that the mortgagor has received the HUD/FHA statement of appraised value required.

(2) A builder's warranty is required on proposed construction.

(3) Eligible for open-end advances where local law permits.

(4) The down payment, settlement costs, and prepaid expenses may be borrowed from an approved individual or corporation.

(5) The VA Certificate of Reasonable Value may be presented to establish the value of property.

Section 203 (k) Loan Insurance For Major Home Improvements

a. *Purpose of Loan:* The purpose of the loan is to finance alteration, repair, or improvement of existing one- to four-family housing.

b. *Amount Insurable:* $12,000 per family unit. Family-unit limits may be increased up to 45% in areas where cost levels so require.

c. *Loan-to-Value Ratio:* The amount of loan plus the existing debt on the property may not exceed the amount insurable under Section 203(b) for the same mortgagor purchasing the same property.

d. *Term of Loan:* 20 years, or ¾ of remaining economic life, whichever is less.

e. *Interest Rate:* Current interest rate as determined by the Secretary of Housing and Urban Development.

f. *Insurance Premium:* ½% on declining scheduled balances.

g. *Initial Service Charge:*

(1) $20 or 1%, whichever is greater, or

(2) $50 or 2½%, whichever is greater, if the lender makes partial disbursements and property inspections while improvements are being completed.

h. *Application Fee:* $20.

i. *Special Factors:*

(1) Housing must be at least ten years old, unless the loan is primarily to make major structural improvements; to correct faults not known when structure was completed, or caused by fire, flood or other casualty; or to construct a civil defense shelter.

(2) The loan proceeds may be used to pay municipal assessments or similar charges for water, sewer, sidewalk, curb, or other public improvements.

(3) An insured loan may be made to the lessee, if the term of the lease will run more than ten years beyond maturity of the loan.

Section 234(c) Home Mortgage Insurance—Condominium Units

a. *Purpose of Mortgage:* The purpose of the mortgage is to finance acquisition of individual units in condominium projects containing four or more units. If the project has more than twelve units, the project mortgage can be insured, as long as the property's construction was finished more than one year before FHA financing was applied for.

b. *Amount Insurable:*

(1) Occupant Mortgagor: $67,500 or $74,500 if cost levels dictate.

(2) Non-Occupant Mortgagor: 85% of occupant mortgagor amount.

c. *Loan-to-Value Ratio:*

(1) Occupant Mortgagor:

97% of first $25,000 of sum of the HUD/FHA estimate of property value and closing costs, plus

90% of next $10,000 of sum of the HUD/FHA estimate of property value and closing costs, plus

80% of sum of the HUD/FHA estimate of property value and closing costs over $35,000.

(2) Non-Occupant Mortgagor: 85% of amount available to occupant mortgagor.

d. *Term of Mortgage:* 30 years or ¾ of remaining economic life, whichever is less. The 30-year limit may be extended to 35 years for an occupant mortgagor unacceptable under the 30-year term.

e. *Interest Rate:* Current interest rate as determined by the Secretary of Housing and Urban Development.

f. *Insurance Premium:* ½% on declining scheduled balances.

g. *Initial Service Charge:* $20 or 1%, whichever is greater.

h. *Application Fee:* $40, except none when the application is filed before issuance of commitment to insure project mortgage.

I. *Special Factors:*

(1) The mortgagor may not own more than four units.

(2) Certification that the mortgagor has received the HUD/FHA statement of appraised value required.

(3) If the mortgagor is 60 years of age or older, downpayment, settlement costs, and prepaid expenses may be borrowed from an approved corporation or individual.

(4) Eligible for open-end advances where local law permits.

Section 234(d) Mortgage Insurance For Condominium Projects

a. *Purpose of Mortgage:* The purpose of the mortgage is to finance construction or rehabilitation of detached, semi-detached, row, walkup, or elevator type housing by a sponsor intending to sell individual units as condominiums. Four or more units are required.

b. *Eligible Mortgagors:* Private and Public Mortgagors.

c. *Maximum Amount Insurable:* Limit per family unit:

(1) Elevator type:

$15,000 no bedroom.

$21,000 one-bedroom.

$25,750 two-bedroom.

$32,250 three-bedroom.

$36,465 four-bedroom or more.

(2) All other types:

$13,000 no bedroom.

$18,000 one-bedroom.

$21,500 two-bedroom.

$26,500 three-bedroom.

$30,000 four-bedroom or more.

(3) In areas where cost levels so require, limits per family unit may be increased up to 45%.

d. *Loan-to-Value Ratio:*

(1) Proposed construction: 90% of the replacement cost or the sum of unit mortgage amounts computed under Section 234(c) for an occupant mortgagor, whichever is less if 80% of the value of the units have been sold prior to initial endorsement, otherwise 80% of the replacement cost.

(2) Rehabilitation: The sum of unit mortgage amounts computed under Section 234(c) for owner-occupant mortgagor subject to following limitations:

(a) Property to be acquired: 90% of the estimated rehabilitation cost plus the lesser of 90% of the purchase price or 90% of the estimated value before rehabilitation.

(b) Property owned:

1) Where property is owned free and clear: 100% of estimated cost of rehabilitation.

2) Where existing indebtedness is to be refinanced with mortgage proceeds: 100% of the estimated rehabilitation cost plus the lesser of existing debt on the property or 90% of the appraised value before rehabilitation.

e. *Term of Mortgage:* Blanket mortgage: 40 years, or not appreciably in excess of ¾ of remaining economic life, whichever is less.

f. *Interest Rate:* Current interest rate as determined by the Secretary of Housing and Urban Development.

g. *Insurance Premium:* ½ % on declining scheduled balances.

h. *Initial Service Charge:* 2%.

i. *Application Fees:*

(1) Application and Commitment:

(a) $1.00 per $1,000 S.A.M.A.

(b) $1.00 per $1,000 Conditional Commitment

(c) $1.00 per $1,000 Firm Commitment

(2) Separate Inspection Fee: As set by HUD/FHA, but not to exceed $5.00 per $1,000.

j. *Special Factors:* Cost certification is required.

Section 213 Home Mortgage Insurance—Sales Type Cooperative

a. *Purpose of Mortgage:* The purpose of the mortgage is to finance an individual mortgage on a dwelling unit released from a cooperative project-sales mortgage.

b. *Amount Insurable Loan-Value Ratio:* The unpaid balance of the project mortgage allocable to individual property.

c. *Term of Mortgage:* 30 years, except 35 years if the mortgagor is unacceptable under a 30-year term.

d. *Interest Rate:* Current interest rate as determined by the Secretary of Housing and Urban Development.

e. *Insurance Premium:* ½% on declining scheduled balances.

f. *Initial Service Charge:* $20 or 1%, whichever is greater.

g. *Application Fee:* None.

h. *Special Factors:*

(1) Eligible for open-end advances where local law permits.

(2) A builder's warranty is required.

Section 240 Purchase of Fee-Simple Title From Lessors

a. *Purpose of Loan:* The purpose of the loan is to finance purchase from lessors by homeowners of fee-simple title to property which is held under long-term ground leases and on which their homes are located.

b. *Amount Insurable:* One- to four-family residences: The cost of purchasing fee-simple title or $10,000 per family unit, whichever is less.

c. *Loan-To-Value-Ratio:* The amount of loan plus the existing debt on the property may not exceed prescribed limits for Section 203(b).

d. *Term of Loan:* 20 years, or ¾ of the remaining economic life of the dwelling, whichever is less.

e. *Interest Rate:* Current interest rate as determined by the Secretary of Housing and Urban Development.

f. *Insurance Premium:* ½% on declining scheduled balances.

g. *Initial Service Charge:* $20 or 1%, whichever is greater.

h. *Application Fee:* $20.

Section 213 Mortgage Insurance For Management-Type Cooperative Project

a. *Purpose of Mortgage:* The purpose of the mortgage is to finance construction, acquisition of existing, or rehabilitation of detached, semi-detached, row, walkup, or elevator type housing by a nonprofit cooperative or acquisition from investor sponsor—five or more units.

b. *Eligible Mortgagors:* Private or Public Mortgagor of a Management Cooperative Type.

c. *Maximum Amount Insurable:* Same as Section 207 Rental Housing Insurance. In areas where cost levels so require, limits per family unit may be increased up to 75%.

d. *Loan-to-Value Ratio:*

(1) Proposed construction: 98% of estimated replacement cost.

(2) Existing construction: 98% of appraised value.

(3) Rehabilitation: 98% of estimated value after rehabilitation.

e. *Term of Mortgage:* 40 years, or not appreciably in excess of ¾ of the remaining economic life, whichever is less.

f. *Interest Rate:* Current interest rate as determined by the

Secretary of Housing and Urban Development.

g. *Insurance Premium:* ½% on declining scheduled balances.

h. *Initial Service Charge:* 2%.

i. *Application Fees:*

(1) Application and Commitment:

(a) $1.00 per $1,000 S.A.M.A.

(b) $1.00 per $1,000 Conditional Commitment

(c) $1.00 per $1,000 Firm Commitment

(2) Separate Inspection Fee: As set by HUD/FHA, but not to exceed $5.00 per $1,000.

Section 213(j) Cooperative Housing Insurance, Management-Type Projects—Supplementary Loans

a. *Purpose of Loan:*

(1) To finance improvement or repair of existing Section 213, management type cooperative housing; construction of community facilities; or resale of cooperative memberships.

(2) To finance rehabilitation or modernization where the property was purchased from the federal government by a cooperative with an uninsured mortgage.

b. *Eligible Mortgagors:* Management Type Cooperative.

c. *Amount Insurable, Loan-Value Ratio:* Estimated costs of repairs or the amount needed to finance sales cannot, when added to the outstanding debt on the property, exceed the original mortgage; a supplementary loan for improvements or for new community facilities can equal 97% of the value of the improvements or new facilities, provided such amount, when added to the other outstanding indebtedness, does not exceed the limit for a Section 213 management mortgage.

d. *Term of Loan:* Remaining term of mortgage. Where property was purchased from the federal government with an uninsured mortgage more than 20 years prior to commitment, the term of the supplementary loan can exceed the remaining term of uninsured mortgage by up to 10 years.

e. *Interest Rate:* Current interest rate as determined by the Secretary of Housing and Urban Development.

f. *Insurance Premium:* ½% on declining scheduled balances.

g. *Initial Service Charge:* 2%.

h. *Application Fees:*

(1) Application and Commitment $3.00 per $1,000

(2) Separate Inspection Fee: As set by HUD/FHA, but not to exceed $5.00 per $1,000.

i. *Special Factors:* Cost certification is required.

Section 213 Cooperative Housing Insurance, Investor-Sponsored Projects

a. *Purpose of Mortgage:* The purpose of the mortgage is to finance construction or rehabilitation of detached, semidetached, row, walkup, or elevator type housing, with five or more units, by a mortgagor intending to sell to a nonprofit cooperative.

b. *Maximum Amount Insurable:* Same as Section 207 Rental Housing Insurance. In areas where cost levels so require, limits per family unit may be increased up to 75%.

c. *Loan-to-Value Ratio:*

(1) Proposed construction: 90% of estimated replacement cost.

(2) Rehabilitation: 90% of appraised value after rehabilitation.

d. *Term of Mortgage:* 40 years, or not appreciably in excess of ¾ of the remaining economic life, whichever is less.

e. *Interest Rate:* Current interest rate as determined by the Secretary of Housing and Urban Development.

f. *Insurance Premium:* ½ % on declining scheduled balances.

g. *Initial Service Charge:* 2%.

h. *Application Fees:*

(1) Application and Commitment:

(a) $1.00 per $1,000 S.A.M.A.

(b) $1.00 per $1,000 Conditional Commitment

(c) $1.00 per $1,000 Firm Commitment

(2) Based upon the commitment, the amount applicable to management project.

(3) Separate Inspection Fee: As set by HUD/FHA, but not to exceed $5.00 per $1,000.

i. *Special Factors:* Cost certification is required.

Section 213 Cooperative Housing Insurance, Sales-Type Projects

a. *Purpose of Mortgage:* The purpose of the mortgage is to finance construction of single-family detached, semi-detached, or row housing for sale to a member of a nonprofit cooperative with five or more units.

b. *Eligible Morgagors:* Private or Public Mortgagor of the Sales Cooperative Type.

c. *Maximum Amount Insurable Loan-Value Ratio:* Sum of separate maximum mortgages on single-family housing insurable for occupant mortgagors under Section 203(b).

d. *Term of Mortgage:* 35 years, or not appreciably in excess of ¾ of the remaining economic life, whichever is less.

e. *Interest Rate:* Current interest rate as determined by the Secretary of Housing and Urban Development.

f. *Insurance Premium:* ½ % on declining scheduled balances.

g. *Initial Service Charge:* 2%.

h. *Application Fees:*

(1) Application and Commitment:

(a) $1.00 per $1,000 S.A.M.A.

(b) $1.00 per $1,000 Conditional Commitment

(c) $1.00 per $1,000 Firm Commitment

(2) Separate Inspection Fee: None.

Section 242 Mortgage Insurance For Hospitals

a. *Purpose of Mortgage:* The purpose of the mortgage is to finance construction or rehabilitation of hospitals, including major movable equipment.

b. *Eligible Mortgagors:* Private non-profit and profit motivated sponsors.

c. *Maximum Amount Insurable:* The mortgage will involve a principal obligation not in excess of 90 percent of the Commissioner's estimate of the replacement cost of the hospital, including the equipment to be used in its operation when the proposed improvements are completed and the equipment is installed.

d. *Loan-to-Value Ratio:*

(1) Proposed construction: 90% of estimated replacement cost.

(2) Rehabilitation:

(a) Property to be acquired: 90% of the estimated rehabilitation cost plus the lesser of 90% of the price or 90% of the estimated value before rehabilitation.

(b) Property owned:

1) 100% of the estimated rehabilitation cost plus the lesser of the existing debt on the property or 90% of the estimated value before rehabilitation.

2) Five times the estimated cost of rehabilitation.

e. *Term of Mortgage:* 25 years.

f. *Interest Rate:* Current interest rate as determined by the Secretary of Housing and Urban Development.

g. *Insurance Premium:* ½ % on declining scheduled balances.

h. *Initial Service Charge:* 2%.

i. *Application Fees:*

(1) Application and Commitment: $3 per $1,000.

(2) Separate Inspection Fee: As set by HUD/FHA, but not to exceed $5 per $1,000.

j. *Special Factors:* A certificate of need from the State agency is required.

Title X Mortgage Insurance For Land Development

a. *Purpose of Mortgage:* The purpose of the mortgage is to finance purchase of land and development of building sites for subdivisions or new communities, including water and sewage systems, streets, etc.

b. *Eligible Mortgagors:* Private Mortgagors.

c. *Loan-to-Value Ratio:* 80% of the estimated value of land before development plus 90% of the estimated cost of development or 85% of the estimated value upon completion, whichever is less.

d. *Term of Mortgage:* Ten years, except longer for separate mortgages for new water and/or sewage systems or for development of a new community.

e. *Interest Rate:* Current interest rate as determined by the Secretary of Housing and Urban Development.

f. *Insurance Premium:*

(1) 2% of the insured mortgage if for three years or less (payable at initial closing).

(2) For mortgages in excess of 3 years, the MIP is based on 2% of the face amount of the mortgage for the first 3 years (payable at initial endorsement) plus 1% of the outstanding principal balance for each year thereafter (payable on the third anniversary date of initial endorsement and on each succeeding anniversary).

(3) For mortgages covering water or sewage systems the insurance premium is 1/12 of 1% of the original face amount of the mortgage for each month or fraction prior to the beginning of amortization. After amortization begins, the annual premium is based on ¾ of 1% of average outstanding principal balance of the mortgage scheduled for the premium year.

g. *Initial Service Charge:* 2%.

h. *Application Fees:*

(1) Application and Commitment: $4.50 per $1,000.

(2) Separate Inspection Fee: None.

i. *Special Factors:* Cost certification is required.

Appendix I

FHA FORM NO. 2019

MARKET SALES SURVEY
ESTIMATES OF MARKET PRICE BY COMPARISON

CONSIDERATIONS	Subject Property	1.			2.			3.			4.			5.		
	Address Data	Address Data	Adjustments −	+	Address Data	Adjustments −	+	Address Data	Adjustments −	+	Address Data	Adjustments −	+	Address Data	Adjustments −	+
Date of Sale																
Sq. Ft. Area																
Number of Rooms																
Room Composition																
Baths																
Construction																
Basement																
Porches, Terraces, Patios																
Garage																
Mechanical Equipment																
Kitchen & Laundry Equip.																
Age of Building																
Location Desirability																
Neighborhood Quality																
Lot Area (Sq. Ft.)																
Street Improvements																
Utilities																
Conditions of Sale																
Design																
Livability																
Condition of Improvements																
Marketability																
Other																
Sale Price																
Total PLUS Adjustments																
Total MINUS Adjustments																
Net Adjustment																
Indicated Market Price for Subject Property																

NOTE: In the adjustments column enter dollar amounts by which subject property varies from comparable properties. If subject is better enter a "Plus" figure and if subject is inferior to the comparable enter a "Minus" amount. The net adjustment will be added or subtracted from sale price of comparable to determine indicated market price for subject property.

190271-P

FHA-Wash., D. C.

Appendix II

LENDING POLICY SURVEY

Lender: ____

Address: ____

Phone: ________ Call ''commitment'' requests to Mr. ________

Lending area: _______________________________________

Loan fees: APR

Origination fee (points) ________ Application fee ________

Lawyer's fee ____________ Other APR charges ____________

Are there additional service charges such as: ______________

Appraisal __________________ Tax Service __________________

Escrow ____________________ Credit Report ________________

Current interest rates: Refinance only Length of loan

Homes to __________ years __________ % ________ years

Homes to __________ years __________ % ________ years

Homes to __________ years __________ % ________ years

OR

__________ % Loan to value ratio (LVR) at __________ % interest

__________ % Loan to value ratio (LVR) at __________ % interest

__________ % Loan to value ratio (LVR) at __________ % interest

Is a termite clearance required? ________

Minimum square feet? ________ Minimum lot size? ________

Prepayment penalty ________ days unearned interest? ________

Any interest reduction privileges? ____________________

Is there an exclusive agent through whom I must deal? ______

Note: Make several copies for your lender's survey.

Appendix III

TABLE OF FACTORS AMORTIZING $1,000

Loan	Interest					
Years	10%	12¼%	13½%	14¼%	14½%	15%
1	87.92	88.97	89.56	89.91	90.03	90.26
5	21.25	22.38	23.01	23.40	23.53	23.79
10	13.22	14.50	15.23	15.68	15.83	16.14
15	10.75	12.17	12.99	13.49	13.66	14.00
20	9.66	11.19	12.08	12.62	12.80	13.17
22	9.39	10.96	11.87	12.43	12.62	12.99
25	9.09	10.72	11.66	12.23	12.43	12.81
27	8.95	10.61	11.56	12.14	12.34	12.73
30	8.78	10.48	11.46	12.05	12.25	12.65
35	8.60	10.36	11.36	11.96	12.17	12.57
40	8.50	10.29	11.31	11.92	12.13	12.54

Multiply the thousands of dollars borrowed by the factor corresponding to the terms in years and the interest rate. Amortization is a direct reduction of the loan with a constant monthly payment which includes principal and interest on the remaining balance (after each payment). No balloon payments are involved in an amortized loan.

EXAMPLES:

Loan of $4,250 for 5 years @ 10%

```
   4.250
  21.250
  ------
   21250
   8500
  4250
 8500
--------
 90.31250
```

or $90.31 per month

Loan of $102,200 for 40 years @15%

```
   102.200
     12.54
   -------
    408800
   511000
  204400
 102200
 ----------
 1281.58800
```

or $1,281.59 per month.

Appendix IV

RATE OF PAY-OFF FORMULA
(Balloon payments)

$$D(A-B) \times [1/2\ (CE) + 1.01]$$

D = Term of note in months
A = Payment necessary to amortize loan
B = Payment stated in note
C = Term of note in years
E = Rate of interest stated in note

Example: A junior lien or 2nd trust deed in the amount of $4,250 is taken back as a purchase money mortgage in lieu of cash. The term of the note in months is 60, the payment necessary to amortize the loan is $90.31. (See Appendix III.) The payment stated in the note is $42.50 (or 1% rate of pay-off). The term of the note in years is 5. The rate of interest stated in the note is 10%.

60(90.31-42.50) x 1/2 (5 x .10) +1.01
60(47.81) x 1/2 (.50) + 1.01
2868.60 x (.25) + 1.01
2868.60 x 1.26 equals $3,614.44

***Balloon payment* $3,614.44 at end of 5 years.**

The above formula is approximate. Accuracy can be determined by reference to the balloon payment balance remaining in 5 years with payments at a rate of 1% per month. This is 87.09 % of the principal amount borrowed, or $86.89 more than the formula above, a 2% difference.

Appendix V

MORTGAGE YIELD GUIDE
(For purchaser of trust deed at discount)

Due in 60 months [five years] at the ordinary rate of pay-off including 10% interest

Percent Yield	Rate of pay-off					Percent Yield
	1.5	1.3	1.2	1.1	1.0*	
10	0	0	0	0	0	10
11	3.1	3.3	3.4	3.6	3.7	11
12	6.0	6.4	6.7	6.9	7.1	12
13	8.7	9.4	9.8	10.1	10.4	13
14	11.5	12.3	12.8	13.2	13.6	14
15	14.0	15.0	15.6	16.1	16.6	15
16	16.5	17.6	18.3	18.9	19.5	16
17	18.9	20.2	20.9	21.6	22.3	17
18	21.2	22.7	23.5	24.2	25.0*	18*
19	23.3	24.9	25.8	26.7	27.5	19
20	25.5	27.2	28.2	29.1	29.9	20
21	27.5	29.4	30.4	31.3	32.3	21
22	29.5	31.4	32.5	33.5	34.5	22

*Indicates 1% of pay-off, 25% discount and an 18% yield.

The discounter who invests in junior mortgages or 2nd trust deeds buys notes that, as a rule, are payable in equal monthly installments which include interest.

The yield guide above represents the return on a discounted investment. When a mortgage is bought at a discount, the investor's yield will be in excess of the interest rate called for by the obligation. The principal amount to be received is considerably more than the money invested.

The difference between a face value of a note and the cost to the investor is the "discount." The amount advanced is the percent of the principal amount due on the note. To compute, the discount is divided by the face value of the note.

Example: **A $4,250 2nd trust deed bought for $3,187.50 equals a discount of $1,062.50. $1,062.50 divided by $4,250 results in a 25% discount. Note above.**

To determine the rate of pay-off, the monthly payment agreed upon is divided by the principal balance as shown in the example:

Monthly payment	Principal balance	Computation	Rate of pay-off
$42.50	$4,250	$42.50 ÷ $4,250	1.0%

10% REMAINING BALLOON BALANCES 10%

(in percent of loan amount)

Due date (Months)	24	36	48	60*	72	84
Rate of Pay-off %						
1%	95.6	93.0	90.2	87.0*	83.6	79.8
2%	69.1	51.2	31.5	9.7	0.0	0.0

* Accuracy dictates 87.09% for 5 years.
(See "formula", APPENDIX IV.)

Appendix VI

REVENUE STAMPS INFORMATION

When the federal government abolished the tax stamp revenue in January, 1968, various states, counties and local jurisdictions instituted a tax on revenue. When the federal government ceased to require the revenue, the charge on all the warranty deeds was based on 55 cents for each 500 dollars of the consideration (cash) or a fraction thereof. These are examples only, not a complete guide.

Consideration	Tax stamps	Consideration	Tax stamps
$ 100 to 500	$ .55	$18,501 to 19,000	$20.90
501 to 1,000	1.10		
1,001 to 1,500	1.65	21,001 to 21,500	23.65
1,501 to 2,000	2.20		
2,001 to 2,500	2.75	24,501 to 25,000	27.50
2,501 to 3,000	3.30		
3,001 to 3,500	3.85	27,001 to 27,500	30.25
10,001 to 10,500	11.55	28,501 to 29,000	31.90
10.501 to 11,000	12.10		
11,001 to 11,500	12.65		
11,501 to 12,000	13.20		
12,001 to 12,500	13.75	29,001 to 29,500	32.45
12,501 to 13,000	14.30		
13,001 to 13,500	14.85		

Although the word "consideration" implies price, tax stamps should not represent the exact amount of deeds of trust for which no revenue stamps are required. Practically speaking, a cash sale of $104,000 should reflect $114.40.

Appendix VII

SCHEDULE OF DIRECT REDUCTION LOAN

Loan $40,400   Rate 7.8%   Payment $320.78   Term 22 yrs.   Period 264

An amortization schedule can be constructed showing each payment into the interest and principal as well as the balance outstanding after each payment has been made.

a) Compute the interest on the previous balance: $40,400 x .078 ÷ 12 (mo's) = $262.60

b) Deduct this from the payment: $320.78 - $262.60 = $58.18

c) Credit the remainder as a repayment of principal: $40,400 - $58.18 = $40,341.82

Payment Number	Net Interest	Principal Payment	Balance of Loan
1	$262.60	$58.18	$40,341.82
2	262.22	58.56	40,283.26
3	261.84	58.94	40,224.32
35	248.26	72.52	38,121.81
36	247.79	72.99	38,048.82
Note:			
60	Complete this for practice		36,147.00

Appendix VIII

TAX EXEMPT INVESTMENT YIELDS COMPARED WITH
TAXABLE INCOME EQUIVALENTS

Your tax bracket	Single taxpayers		Tax exempt yields of			Joint taxpayers		Your tax bracket
	Over	Not over	7.25%	8.50%	10.00%	Over	Not over	
			Compares to taxable yield of					
26%	$12,900 -	$15,000	9.80%	11.49%	13.51%			
			9.54%	11.18%	13.16%	$16,000 -	$20,200	24%
30%	$15,000 -	$18,200	10.36%	12.14%	14.29%			
			10.07%	11.81%	13.89%	$20,200 -	$24,600	28%
34%	$18,200 -	$23,500	10.98%	12.88%	15.15%			
			10.66%	12.50%	14.71%	$24,600 -	$29,900	32%
39%	$23,500 -	$28,800	11.89%	13.93%	16.39%			
			11.51%	13.49%	15.87%	$29,900 -	$35,200	37%
44%	$28,800 -	$34,100	12.95%	15.89%	17.86%			
			12.72%	14.91%	17.54%	$35,200 -	$45,800	43%
49%	$34,100 -	$41,500	14.22%	16.67%	19.61%			
			14.22%	16.67%	19.61%	$45,800 -	$60,000	49%
55%	$41,500 -	$55,300	16.11%	18.89%	22.22%			
			15.76%	18.48%	21.74%	$60,000 -	$85,600	54%
63%	$55,300 -	$81,800	19.60%	22.97%	27.03%			
			17.68%	20.73%	24.39%	$85,600 -	$109,400	59%
68%	$81,800 -	$108,300	22.66%	26.56%	31.25%			
			20.14%	23.61%	27.78%	$109,400 -	$162,400	64%
70%	$108,300 -	------	24.17%	28.33%	33.33%			
			22.66%	26.56%	31.25%	$162,400 -	$215,400	68%

Federal tax brackets used in the table have been taken from the IRS tax tables found in Instruction 1040, used in preparing 1980 returns. Deductions and exemptions must be subtracted from your gross income in order to determine your bracket. Example: A $20,000 TD or mortgage investment earning 10.98 percent interest is subject to federal income tax at 34 percent of your net taxable yearly income is, say $23,500 as a single taxpayer. You actually net $2,196 minus 34 percent if this amount, or $1,450. This is equivalent to 7.25 percent earned from a tax free investment such as municipal bonds, or $20,000 multiplied by .0725 equals $1,450 in which you may pocket the interest dollars earned. However, if your cash investments are *proceeds* of interest bearing mortgage *refinance* loan subject to interest deduction from gross income, you must calculate from the table of taxable yields.

In comparing yields from a single family building project

Appendix IX

FORECLOSURE METHODS IN THE UNITED STATES AND POSSESSIONS

	Mortgage theory	Security device	Remedy process	Redemption term	Interim holder
Alabama	Title	Mortgage	Power of sale	12 months	Purchaser
Alaska	Lien	Trust deed	Power of sale	12 months	Mortgagor
Arizona	Lien	Mortgage	Court action	6 months	Mortgagor
Arkansas	Medial	Mortgage	Court action	None (1)	Absolved
California	Lien	Trust deed	Power of sale	None	Absolved
Colorado	Lien	Trust deed	Power of sale	6 months	Mortgagor
Connecticut	Medial	Mortgage	Strict foreclosure	None	Absolved
Delaware	Medial	Mortgage	Court action	None	Absolved
Florida	Lien	Mortgage	Court action (8)	None	Absolved
Georgia	Title	Security deed	Power of sale	None	Absolved
Hawaii	Lien	Mortgage	Court action	12 months	Mortgagor
Idaho	Lien	Trust deed	Power of sale	12 months (2)	Mortgagor
Illinois	Medial	Trust deed	Court action	12 months (10)	Mortgagor
Indiana	Lien	Mortgage	Court action	12 months	Mortgagor
Iowa	Lien	Mortgage	Court action	12 months	Mortgagor
Kansas	Lien	Mortgage	Court action	12 months	Mortgagor
Kentucky	Lien	Mortgage	Court action	12 months (3)	Mortgagor
Louisiana	Lien	Mortgage	Court action	None	Absolved
Maine	Title	Mortgage	Public notice	12 months	Mortgagor
Maryland	Title	Mortgage	Power of sale (4)	None	Absolved
Massachusetts	Medial	Mortgage	Power of sale	None	Absolved
Michigan	Lien	Mortgage	Power of sale	12 months (5)	Mortgagor
Minnesota	Lien	Mortgage	Power of sale	6 months (6)	Mortgagor
Mississippi	Medial	Trust deed	Power of sale	None	Absolved
Missouri	Medial	Trust deed	Power of sale	12 months (7)	Mortgagor
Montana	Lien	Mortgage	Court action	12 months	Mortgagor
Nebraska	Lien	Mortgage	Court action	9 months	Mortgagor
Nevada	Lien	Mortgage	Court action	12 months	Mortgagor
N. Hampshire	Title	Mortgage	Power of sale	None	Absolved
N. Jersey	Medial	Mortgage	Court action	None (9)	Absolved
New Mexico	Lien	Mortgage	Court action	9 months (11)	Purchaser
New York	Lien	Mortgage	Court action	None	Absolved
N. Carolina	Medial	Trust deed	Power of sale	None	Absolved
N. Dakota	Lien	Mortgage	Court action	12 months	Mortgagor
Ohio	Lien	Mortgage	Court action	None	Absolved
Oklahoma	Lien	Mortgage	Court action	None	Absolved
Oregon	Lien	Mortgage	Court action	12 months	Purchaser
Pennsylvania	Title	Mortgage	Court action	None	Absolved
Rhode Island	Title	Mortgage	Power of sale	None	Absolved
S. Carolina	Lien	Mortgage	Court action	None	Absolved
S. Dakota	Lien	Mortgage	Power of sale	12 months	Mortgagor
Tennessee	Title	Trust deed	Power of sale	None (12)	Absolved

versus yields from tax exempt securities (*or taxable* investments) study the market for the complete project in anticipation of final profit.

	Mortgage theory	Security device	Remedy process	Redemption term	Interim holder
Texas	Lien	Trust deed	Power of sale	None	Absolved
Utah	Lien	Mortgage	Court action	6 months	Mortgagor
Vermont	Title	Mortgage	Strict foreclosure	6 months (13)	Mortgagor
Virginia	Medial	Trust deed	Power of sale	None	Absolved
Washington	Lien	Mortgage	Court action	12 months	Purchaser
W. Virginia	Medial	Trust deed	Power of sale	None	Absolved
Wisconsin	Lien	Mortgage	Court action	None	Absolved
Wyoming	Lien	Mortgage	Power of sale	3 months (14)	Mortgagor

District of Columbia mortgage foreclosure theory favors a title with trust deeds as a security device, power of sale and no redemption term.

Puerto Rico adopts the lien theory with the mortgage as security device. Foreclosure is by court action with no redemption period upon order of sale.

Virgin Islands. The lien theory prevails with the mortgage as the security device. Foreclosures are by court order providing a 6 month redemption term during which the mortgagor retains possession.

EXPLANATORY NOTES

(1) The courts of Arkansas have not consistently adopted mortgage theory as a lien or as title, since some dispositions have favored both. Where power of sale prevails in a mortgage, the mortgagor may redeem within one year.

(2) Redemption in Idaho is limited to a 6 month period if the property consists of 20 acres or less.

(3) Mortgagors in Kentucky may redeem in 12 months time if

the court ordered sale produces ⅔ or less of court ordered appraisal.

(4) In Maryland the power of sale in a mortgage is subject to the supervision by the Court of Equity of the county, or city of Baltimore when property lies within city.

(5) Foreclosures by court action in Michigan reduce redemption period to 6 months. The one year redemption period under a power of sale contained in a mortgage is subject to several variations of realty classifications.

(6) Mortgagors in Minnesota have 12 months to redeem if the mortgage was executed before July 1, 1967. Other details may qualify under special study.

(7) A redemption notice together with costs to be incurred must be made in Missouri 10 days prior to the sale in order to qualify for 12 months redemption rule.

(8) No redemption after a court ordered sale. Trust deeds in the state of Florida can be a security device creating a lien but legal title does not pass until property is sold.

(9) An exception in New Jersey provides an objection period of ten days set aside for redemption after the sale.

(10) In Illinois redemption is limited to 60 days in cases of abandonment.

(11) New Mexico courts may extend redemption period up to 9 months (upon special showing). If a mortgage specifies a short redemption period the legal limit is 30 days.

(12) Provided the redemption privilege is waived in the trust deed. When no waiver appears, Tennessee permits a 2 year period for redemption.

(13) Prior to April 1, 1968 the mortgage redemption period is 1 year. Residence vesting is absolute in mortgagee when redemption lost.

(14) Wyoming agricultural mortgagors may redeem as much as 9 months from sale date, or to November 1 of sale year if greater term.

Appendix X

REAL PROPERTY TAX RATES AND ASSESSED VALUES IN GOVERNMENT REVENUE

A governmental unit establishes the total expenses for the coming year for public services in its jurisdiction. It determines the tax rate by dividing the total revenue to be raised from tax by the total taxable assessed value of property.

Thus, if a county estimates its revenue from property tax at $2,875,500 and the taxable property has a total value of $100,500,000, as assessed, the tax rate of the property if all assessed value is taxable would be as follows:

$$\text{Tax rate equals } \frac{\text{Estimated revenue from tax}}{\text{Taxable assessed value}} \text{ equals } \frac{\$\ 2{,}875{,}500}{\$100{,}500{,}000}$$

or .028611 per dollar of assessed value of property.

Tax rates may also be expressed as 28.611 mills per dollar or 2.8611% of the assessed value of the property ($1 equalling 1000 mills).

The tax rate may be based upon only 80% of the assessed value. With the same amount of revenue to be obtained from 80% of assessed value the tax rate would be determined as follows:

$$\text{Tax rate equals } \frac{\text{Estimated revenue from tax}}{\text{Taxable assessed value}} \text{ equals } \frac{\$\ 2{,}875{,}500}{\$100{,}500{,}000 \times 80\%}$$

equals .03576 or .03576 per dollar of 80% of assessed value. Or it can be expressed as 35.76 mills per dollar or 3.57% of 80% of the assessed value.

Example: If your property is assessed at $5,000 in your county and the tax rate is 100% of assessed valuation (as in the first example) then $5,000 x .028611 equal $143.05 (tax owed). But if your property is assessed at $5,000 in your county and the tax rate is 80% of assessed valuation, then $5,000 x 80% x

.035764 equals $143.05. The taxable value is $5,000 x 80%, or $4,000. The tax owed is $143.05.

The same general principles of assessment are followed by cities and counties throughout the land; however, some cities make their own independent assessed valuations. The city council performs the same functions as the county board of supervisors. Some incorporated cities arrange with the county to handle all city tax assessments and the collections.

If a city has a property tax rate of 85 mills per dollar on 60% of the assessed value and your property assessment is $40,000, what is your tax? Since the taxable value equals $40,000 x 60% or $24,000 then the multiplier will be in mills (or .085). The tax is $24,000 times 8½% or $2,040.

(a) Find the tax rate in mills to two decimal places, and (b) find the tax on the individual properties listed below.

Revenue required by taxing	Total assessed value in county	Assessed value each property	Taxable value
$25,000	$ 800,000	$200.	100%
50,000	3,500,000	4000.	70%
83,400	6,520,000	1000.	90%
185,000	8,520,000	500.	85%

Federal Taxes

Although not directly of interest to borrowers or lenders of money secured by mortgages or trust deeds, it is appropriate to mention federal estate taxes at this time:

The United States imposes a tax on the net estate of a decedent when the estate exceeds certain amounts. This is not an inheritance tax, and is not computed on the amount of each inheritance. Unless sooner paid in full, the tax is a lien for ten years upon the gross estate of the decedent, except such a part of it which might be used for the payment of charges against the estate and expenses of administration allowed by any court having jurisdiction. Otherwise the lien follows the property into the possession of distributees or purchasers.

The Federal Estate Tax is based on property values of both lifetime gifts and testimentary transfers of title. The 1976 Tax Reform Act unified the above transfer taxes on both types of transfers so that a single tax will be paid on unified tax rates.

The amount of the unified credit allowed the estates of individuals dying in 1977 is $30,000. The credit is increased as follows:

For individuals dying in:	Credit
1978	$34,000
1979	$38,000
1980	$42,500
1981	$47,000

The unified tentative transfer tax rates range from 18% in amounts not over $10,000 to 70% for gifts or estates over the amount of $5,000,000. A marital deduction is permitted for estates with certain qualifying property left to the surviving spouse; a deduction being one-half of the adjusted gross estate or $250,000 whichever is greater.

Example: A decedent left an estate of $300,000 to a surviving spouse in the year 1980. The marital deduction of $250,000 leaves $50,000 for computing tax. Since the amount for which the tentative tax is to be computed is only $10,600, the unified credit of $42,500 (for individuals dying in 1980) clearly eliminates a Federal Estate Tax. However, if the gift tax credit had been previously used to reduce the tax on lifetime gift transfers the unified credit of $42,500 above, would be affected by further reduction.

The Unified Tax Rate which is the same as for estate taxes is shown in the tax rate schedules of Publication No. 448 of the Internal Revenue Service, titled *A Guide To Federal Estate and Gift Taxation, 1979 Edition* available from the Superintendent of Documents, U.S. Government Printing Office, Washington D.C. 20402. Stock Number 048-004-01806-3. Cost $3.75.

Appendix XI

CONSTRUCTION FINANCING RISK AND RETURN YIELDS

Table A is based on a $100,000, 6-month construction loan bearing 12.00% interest with 1 point (1% origination fee) paid in advance. Lender's return yield is "median interest." Installment delay reduces median rate.

Table A

Monthly disbursement	Loan balance each month	This month interest	All interest plus 1 point ($1,000)	Median rate of interest
1. $16,666	16,666	166.66		
2. $16,666	33,332	333.32		
3. $16,666	49,998	499.98		
4. $16,666	66,664	666.64		
5. $16,666	83,330	833.30		
6. $16,670	100,000	1,000.00		
$100,000	$350,000	$3,500.00	$4,500	15.43%

If delay occurs in repayment to 7th month, median rate will be 14.67%

If delay occurs in repayment to 8th month, median rate will be 14.18%

Table B

Table B is based on an $80,000, 6-month construction loan bearing 10.00% interest with 3 points ($2,400.) from loan proceeds in advance. As noted above, the return yield on original term is based upon "median interest." Payment past loan period reduces this median interest rate.

Monthly disbursement	Loan balance each month	This month interest	All interest plus 3 points ($2,400)	Median rate of interest
1. $13,333	13,333	111.11		
2. $13.333	26,666	222.22		

Table B (continued)

Monthly disbursement	Loan balance each month	This month interest	All interest plus 3 points ($2,400)	Median rate of interest
3. $13,333	39,999	333.33		
4. $13,333	53,332	444.43		
5. $13,333	66,665	555.54		
6. $13,335	80,000	666,65		
$80,000	$280,000	$2,333.28	$4,733.28	20.29%

If delay occurs in repayment to 7th month, median rate will be 18.00%

Note: To get lender's median rate based on original loan period, divide *all interest* plus points by *total loan balance,* then multiply by 12. Example, Table A:

$4,500 ÷ $350,000 x 12 = .1542852 (6th mo.)
$5,500 ÷ $450,000 x 12 = .1466664 (7th mo.)

Appendix XII

LAND DESCRIPTION CONVERSION GUIDE

A section of land - 640 acres

A rod is 16½ feet

A chain is 66 feet or 4 rods

A mile is 320 rods, 80 chains, or 5,280 feet

A square rod is 272¼ square feet

An acre contains 43,560 square feet

An acre contains 160 square rods

An acre is about 208¾ feet square

An acre is 8 rods wide by 20 rods long or any two numbers (of rods) whose product is 160

25 x 125 feet = .0717 of an acre

Center

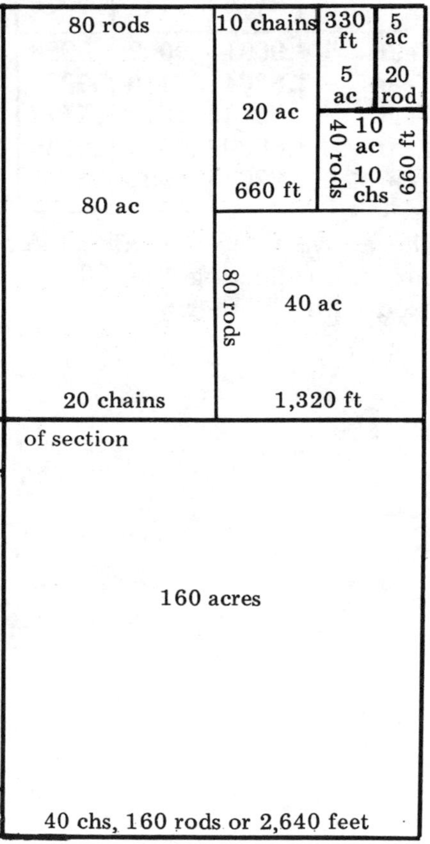

80 rods

80 ac

20 chains

10 chains

20 ac

660 ft

330 ft

5 ac

5 ac

20 rod

10 ac

40 rods

10 chs

660 ft

80 rods

40 ac

1,320 ft

of section

160 acres

40 chs, 160 rods or 2,640 feet

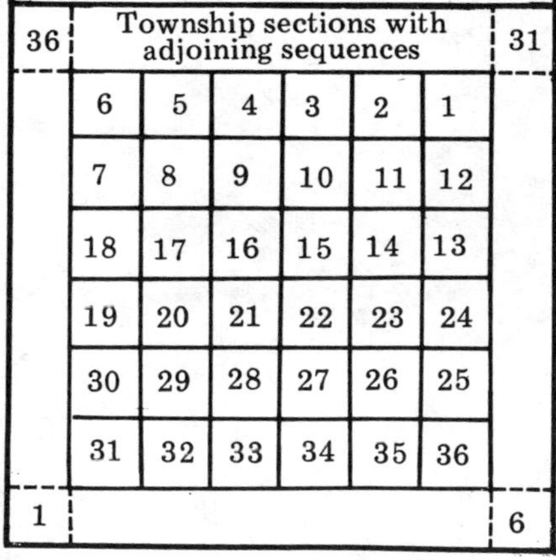

Township sections with adjoining sequences

36							31
	6	5	4	3	2	1	
	7	8	9	10	11	12	
	18	17	16	15	14	13	
	19	20	21	22	23	24	
	30	29	28	27	26	25	
	31	32	33	34	35	36	
1							6

Appendix XIII

DISCOUNT FACTORS IN TRACT FINANCING

INTEREST RATES

	10%	11%	12%	13%	14%	15%	
Years 1	.9091	.9009	.8928	.8849	.8771	.8695	1 Years
Years 2	.8264	.8116	.7971	.7831	.7694	.7561	2 Years
Years 3	.7513	.7311	.7117	.6930	.6749	.6575	3 Years
Years 4	.6830	.6587	.6355	.6133	.5920	.5717	4 Years
Years 5	.6209	.5934	.5674	.5427	.5193	.4971	5 Years

The above table indicates the single investment today, with interest, reflecting one dollar in one to five years' time. To be used with Chapter 9.

Appendix XIV

SELECTED PAMPHLETS FOR BUILDER'S LIBRARY

HOUSING AND URBAN DEVELOPMENT (HUD/FHA)

Digest of Insurable Loans 4000.1
Mortgagee's Guide—Application through Firm Commitment 4000.2
HUD/FHA Underwriting Analysis with Changes 4020.1
Construction Complaints & Section 518(a) and (b), with changes 4070.1
General Program Summary for Section 203(b) for Home Mortgage Insurance with changes 4100.1
Underwriting—Technical Direction for Home Mortgage Insurance with changes 4125.1
Cost Estimation/Cost Data for Home Mortgage Insurance with changes 4130.1
Procedures for Approval of S.F. Proposed Construction Applications in New Subdivisions 4135.1 REV-1
Land Planning Principles for Home Mortgage Insurance 4140.1
Architectural Processing and Inspection for Home Mortgage Insurance—Section 203(b) with changes 4145.1
Architectural Handbook for Building Single Family Dwellings for Home Mortgage Insurance 4145.2
Valuation Analysis for Home Mortgage Insurance with changes 4150.1
Mortgage Credit Analysis Handbook for Mortgage Insurance on One to Four-Family Properties with changes 4155.1
Reconsideration (Before Endorsement) for Home Mortgage Insurance 4160.1
Endorsement for Insurance for Home Mortgage Programs with changes 4165.1
Reconsideration (After Endorsement) for Home Mortgage Insurance with changes 4170.1
Single Family Underwriting Reports and Forms Catalog 4190.1
Homeownership for Lower Income Families 4210.1

Section 203(h) Home Mortgage Insurance for Disaster Victims, Section 203(i) Home Mortgage Insurance for Outlying Areas, Section 203(k) Loan Insurance with Major Home Improvements	4240.1
Section 220(d) (3) (A) and Section 220(h), Rehabilitation with changes	4245.1
Home Mortgage Insurance for Low or Moderate Income-Section 221(d) (2)	4250.1
Home Mortgage Insurance—Condominium Units—Section 234(c) with changes, and Section 234(d), (Multifamily)	4265.1
Home Mortgage Insurance—Sales Type Cooperative—Section 213 (Homes)	4550.6
Purchase of Fee Simple Title from Lessors, Section 240, (Homes)	4270.1
Mortgage Insurance for Management-type Cooperative Project, Section 213 (Multifamily)	4550.2
Cooperative Housing Insurance, Management-Type Projects—Supplementary Loans, Section 213(j)	4550.4
Cooperative Housing Insurance, Investor-Sponsored Projects, Section 213	4550.1
Cooperative Housing Insurance, Sales-Type Projects, Section 213	4550.1
Mortgage Insurance for Hospitals, Section 242 (Multifamily)	4615.1
Mortgage Insurance for Land Development, (Other Programs)	4140.2
Graduated Mortgage Program, Section 245	4240.2
Privacy Act Handbook	1325.1
Freedom of Information	1327.1

VETERANS ADMINISTRATION

Lender's Handbook (revised), December 1979. Available directly from regional VA offices	VA Pamphlet 26-7

Appendix XV

FIELD OFFICE JURISDICTIONS OF HUD

REGION I

Regional Administrator
Rm. 800, John F. Kennedy
Federal Building
Boston, Massachusetts 02203
Tel. (617) 223-4066

AREA OFFICES

CONNECTICUT, HARTFORD 06105
999 Asylum Avenue
Tel. (203) 244-3638

MASSACHUSETTS, BOSTON 02114
Bulfinch Building
15 New Chardon Street
Tel. (617) 233-4111

NEW HAMPSHIRE, MANCHESTER 03101
Davison Building
1230 Elm Street
Tel. (603) 669-7681

INSURING OFFICES

MAINE, BANGOR 04401
Federal Building and Post Office
202 Harlow Street
Post Office Box 1357
Tel. (207) 942-8271

RHODE ISLAND, PROVIDENCE 02903
300 Post Office Annex
Tel. (401) 528-4351

VERMONT, BURLINGTON 05401
Federal Building
Elmwood Avenue
Post Office Box 989
Tel. (802) 862-6501

REGION II

Regional Administrator
26 Federal Plaza, Room 3541
New York, New York 10007
Tel. (212) 264-8068

AREA OFFICES

NEW JERSEY, CAMDEN 08103
The Parkade Building
519 Federal Street
Tel. (609) 963-2541

NEW JERSEY, NEWARK 07102
Gateway 1 Building
Raymond Plaza
Tel. (201) 645-3010

NEW YORK, BUFFALO 14202
Grant Building
560 Main Street
Tel. (716) 842-3510

NEW YORK, NEW YORK 10019
666 Fifth Avenue
Tel. (212) 974-6800

COMMONWEALTH AREA OFFICE

PUERTO RICO, SAN JUAN 00936
Post Office Box 3869 GPO
255 Ponce de Leon Avenue
Hato Rey, Puerto Rico
Tel. (809) 765-0404

INSURING OFFICES

NEW YORK, ALBANY 12206
Westgate North
30 Russell Road
Tel. (518) 472-3567

REGION III

Regional Administrator
Curtis Building
6th and Walnut Streets
Philadelphia, Pennsylvania 19106
Tel. (215) 597-2560

AREA OFFICES

DISTRICT OF COLUMBIA, WASHINGTON 20009
Universal North Building
1875 Connecticut Ave. N.W.
Tel. (202) 382-4855

MARYLAND, BALTIMORE 21201
Two Hopkins Plaza
Mercantile Bank and Trust Building
Tel. (301) 962-2121

PENNSYLVANIA, PHILADELPHIA 19106
Curtis Building
625 Walnut Street
Tel. (215) 597-2665

PENNSYLVANIA, PITTSBURGH 15212
Two Allegheny Center
Tel. (412) 644-2802

VIRGINIA, RICHMOND 23219
701 East Franklin Street
Tel. (804) 782-2721

INSURING OFFICES

DELAWARE, WILMINGTON 19801
Farmers Bank Building
919 Market Street, 14th Floor
Tel. (302) 571-6330

WEST VIRGINIA, CHARLESTON 25330
New Federal Building
500 Quarrier Street
Post Office Box 2948
Tel. (304) 343-6181

REGION IV

Regional Administrator
Room 211, Pershing Point Plaza
1371 Peachtree Street, N.E.
Atlanta, Georgia 30309
Tel. (404) 526-5585

AREA OFFICES

ALABAMA, BIRMINGHAM 35233
Daniel Building
15 South 20th Street
Tel. (205) 325-3264

FLORIDA, JACKSONVILLE 32204
Peninsular Plaza
661 Riverside Avenue
Tel (904) 791-2626

GEORGIA, ATLANTA 30303
Peachtree Center Building
230 Peachtree Street, N.W.
Tel. (404) 526-4576

KENTUCKY, LOUISVILLE 40201
Children's Hospital Foundation Bldg.
601 South Floyd Street
Post Office Box 1044
Tel. (502) 582-5251

MISSISSIPPI, JACKSON 39213
101-C Third Floor Jackson Mall
300 Woodrow Wilson Avenue, W.
Tel. (601) 969-4703

NORTH CAROLINA, GREENSBORO 27408
2309 West Cone Boulevard
Northwest Plaza
Tel. (919) 275-9111

SOUTH CAROLINA, COLUMBIA 2920
1801 Main Street
Jefferson Square
Tel. (803) 765-5591

TENNESSEE, KNOXVILLE 37919
One Northshore Building
1111 Northshore Drive
Tel. (615) 524-1222

INSURING OFFICES

FLORIDA, CORAL GABLES 33134
3001 Ponce de Leon Boulevard
Tel. (305) 445-2561

FLORIDA, TAMPA 33679
4224-28 Henderson Boulevard
Post Office Box 18165
Tel. (813) 228-2501

TENNESSEE, MEMPHIS 38103
28th Floor, 100 North Main Street
Tel. (901) 534-3143

TENNESSEE, NASHVILLE 37203
1717 West End Building
Tel. (615) 749-5521

REGION V

Regional Administrator
300 South Wacker Drive
Chicago, Illinois 60606
Tel. (312) 353-5680

AREA OFFICES

ILLINOIS, CHICAGO 60602
17 North Dearborn Street
Tel. (312) 353-7660

INDIANA, INDIANAPOLIS 46205
Willowbrook 5 Building
4720 Kingsway Drive
Tel. (317) 633-7188

MICHIGAN, DETROIT 48226
5th Floor, First National Building
660 Woodward Avenue
Tel. (313) 226-7900

MINNESOTA, MINNEAPOLIS - ST PAUL
Griggs-Midway Building
1821 University Avenue
St. Paul, Minnesota 55104
Tel. (612) 725-4701

OHIO, COLUMBUS 43215
60 East Main Street
Tel. (614) 469-7345

WISCONSIN, MILWAUKEE 53203
744 North 4th Street
Tel. (414) 224-3223

INSURING OFFICES

ILLINOIS SPRINGFIELD 62704
Lincoln Tower Plaza
524 South Second Street, Room 600
Tel. (217) 525-4414

MICHIGAN, GRAND RAPIDS 49505
Northbrook Building Number II
2922 Fuller Avenue, N.E.
Tel. (616) 456-2225

OHIO, CINCINNATI 45202
Federal Office Building
550 Main Street, Room 9009
Tel. (513) 684-2884

OHIO, CLEVELAND 44114
777 Rockwell
Tel. (216) 522-4065

REGION VI

Regional Administrator
Room 14C2, New Dallas Federal Building
1100 Commerce Street
Dallas, Texas 75202
Tel. (214) 749-7401

AREA OFFICES

ARKANSAS, LITTLE ROCK 72201
Room 1490, Union National Plaza
Tel. (501) 378-5401

LOUISIANA, NEW ORLEANS 70113
Plaza Tower
1001 Howard Avenue
Tel. (504) 527-2063

OKLAHOMA, OKLAHOMA CITY 73102
301 North Hudson Street
Tel. (405) 231-4891

TEXAS, DALLAS 75201
2001 Bryan Tower, 4th Floor
Tel. (214) 749-1601

TEXAS, SAN ANTONIO 78285
Kallison Building
410 South Main Avenue
Post Office Box 9163
Tel. (512) 225-5511

INSURING OFFICES

LOUISIANA, SHREVEPORT 71120
New Federal Building
500 Fannin, 6th Floor
Tel. (318) 425-1241

NEW MEXICO, ALBUQUERQUE 87110
625 Truman Street, N.E.
Tel. (505) 766-3251

OKLAHOMA, TULSA 74152
1708 Utica Square
Post Office Box 4054
Tel. (918) 581-7435

TEXAS, FORT WORTH 76102
819 Taylor Street
Room 13A01 Federal Building
Tel. (817) 334-3233

TEXAS, HOUSTON 77046
Two Greenway Plaza East, Suite 200
Tel. (713) 226-4335

TEXAS, LUBBOCK 79408
Courthouse and Federal Office Building
1205 Texas Avenue
Post Office Box 1647
Tel. (806) 762-7265

REGION VII

Regional Administrator
Federal Office Building, Room 300
911 Walnut Street
Kansas City, Missouri 64106
Tel. (816) 374-2661

AREA OFFICES

KANSAS, KANSAS CITY 66101
Two Gateway Center
4th and State Streets
Tel. (816) 374-4355

MISSOURI, ST. LOUIS 63101
210 North 12th Street
Tel. (314) 622-4761

NEBRASKA, OMAHA 68106
Univac Building
7100 West Center Road
Tel. (402) 221-9301

INSURING OFFICES

IOWA, DES MOINES 50309
210 Walnut Street
Room 259 Federal Building
Tel. (515) 284-4512

KANSAS, TOPEKA 66603
700 Kansas Avenue
Tel. (913) 234-8241

REGION VIII

Regional Administrator
Federal Building
1961 Stout Street
Denver, Colorado 80202
Tel. (303) 837-2741

INSURING OFFICES

COLORADO, DENVER 80202
4th Floor, Title Building
909-17th Street
Tel. (303) 837-2441

MONTANA, HELENA 59601
616 Helena Avenue
Tel. (406) 442-3237

NORTH DAKOTA, FARGO 58102
Federal Building
653-2nd Avenue N.
Post Office Box 2483
Tel. (701) 237-5136

SOUTH DAKOTA, SIOUX FALLS 57102
119 Federal Building U. S. Courthouse
400 S. Phillips Avenue
Tel. (605) 336-2980

UTAH, SALT LAKE CITY 84111
125 South State Street
Post Office Box 11009
Tel. (801) 524-5237

WYOMING, CASPER 82601
Federal Office Building
100 East B Street
Post Office Box 580
Tel. (307) 265-5550

REGION IX

Regional Administrator
450 Golden Gate Avenue
Post Office Box 36003
San Francisco, California 94102
Tel. (415) 556-4752

AREA OFFICES

CALIFORNIA, LOS ANGELES 90057
2500 Wilshire Boulevard
Tel. (213) 688-5973

CALIFORNIA, SAN FRANCISCO 94111
1 Embarcadero Center
Suite 1600
Tel. (415) 556-2238

INSURING OFFICES

ARIZONA, PHOENIX 85002
244 West Osborn Road
Post Office Box 13468
Tel. (602) 261-4435

CALIFORNIA, SACRAMENTO 95809
801 I Street
Post Office Box 1978
Tel. (916) 449-3471

CALIFORNIA, SAN DIEGO 92112
110 West C Street
Post Office Box 2648
Tel. (714) 293-5310

CALIFORNIA, SANTA ANA 92701
1440 East First Street
Tel. (714) 836-2451

HAWAII, HONOLULU 96813
1000 Bishop Street, 10th Floor
Post Office Box 3377
Tel. (808) 546-2136

NEVADA, RENO 89505
1050 Bible Way
Post Office Box 4700
Tel. (702) 784-5356

REGION X

Regional Administrator
Arcade Plaza Building
1321 Second Avenue
Seattle, Washington 98101
Tel. (206) 442-5415

AREA OFFICES

OREGON, PORTLAND 97204
520 Southwest 6th Avenue
Tel. (503) 221-2561

WASHINGTON, SEATTLE 98101
Arcade Plaza Building
1321 Second Avenue
Tel. (206) 442-7456

INSURING OFFICES

ALASKA, ANCHORAGE 99501
334 West 5th Avenue
Tel. (907) 272-5561 Ext. 791

IDAHO, BOISE 83707
331 Idaho Street
Post Office Box 32
Tel. (208) 342-2711

WASHINGTON, SPOKANE 99201
West 920 Riverside Avenue
Tel. (509) 456-2510

Index

Other Useful References

Builder's Guide to Government Loans
This comprehensive guide will help you take advantage of the many government loan programs. Everything is explained in step-by-step instruction and actual sample forms are included: HUD and FHA loan insurance and guarantees for operative builders, interest subsidies, rent supplements, public housing programs, turnkey projects, VA loans, FmHA rural housing programs, and Small Business Administration loans for builders.
416 pages, 5½ x 8½, $13.75

National Construction Estimator
Current building costs in dollars and cents for residential, commercial and industrial construction. Prices for every commonly used building material, and the proper labor cost associated with installation of the material. Everything figured out to give you the "in place" cost in seconds. Many time-saving rules of thumb, waste and coverage factors and estimating tables are included.
512 pages, 8½ x 11, $16.00. Revised annually.

Building Cost Manual
Square foot costs for residential, commercial, industrial, and farm buildings. In a few minutes you work up a reliable budget estimate based on the actual materials and design features, area, shape, wall height, number of floors and support requirements. Most important, you include all the important variables that can make any building unique from a cost standpoint.
240 pages, 8½ x 11, $12.00. Revised annually

Estimating Home Building Costs
Estimate every phase of residential construction from site costs to the profit margin you should include in your bid. Shows how to keep track of manhours and make accurate labor cost estimates for footings, foundations, framing and sheathing finishes, electrical, plumbing and more. Explains the work being estimated and provides sample cost estimate worksheets with complete instructions for each job phase.
320 pages, 5½ x 8½, $14.00

Cost Records for Construction Estimating
How to organize and use cost information from jobs just completed to make more accurate estimates in the future. Explains how to keep the cost records you need to reflect the time spent on each part of the job. Shows the best way to track costs for sitework, footing, foundations, framing, interior finish, siding and trim, masonry, and subcontract expense. Provides sample forms.
208 pages, 8½ x 11, $15.75

Rough Carpentry
All rough carpentry is covered in detail: sills, girders, columns, joists, sheathing, ceiling, roof and wall framing, roof trusses, dormers, bay windows, furring and grounds, stairs and insulation. Many of the 24 chapters explain practical code approved methods for saving lumber and time without sacrificing quality. Chapters on columns, headers, rafters, joists and girders show how to use simple engineering principles to select the right lumber dimension for whatever species and grade you are using.
288 pages, 8½ x 11, $14.50
